Water Science and Technology Library

Volume 89

The aim of the *Water Science and Technology Library* is to provide a forum for dissemination of the state-of-the-art of topics of current interest in the area of water science and technology. This is accomplished through publication of reference books and monographs, authored or edited. Occasionally also proceedings volumes are accepted for publication in the series. *Water Science and Technology Library* encompasses a wide range of topics dealing with science as well as socio-economic aspects of water, environment, and ecology. Both the water quantity and quality issues are relevant and are embraced by *Water Science and Technology Library*. The emphasis may be on either the scientific content, or techniques of solution, or both. There is increasing emphasis these days on processes and *Water Science and Technology Library* is committed to promoting this emphasis by publishing books emphasizing scientific discussions of physical, chemical, and/or biological aspects of water resources. Likewise, current or emerging solution techniques receive high priority. Interdisciplinary coverage is encouraged. Case studies contributing to our knowledge of water science and technology are also embraced by the series. Innovative ideas and novel techniques are of particular interest.

Comments or suggestions for future volumes are welcomed.

Vijay P. Singh, Department of Biological and Agricultural Engineering & Zachry Department of Civil Engineering, Texas A and M University, USA
Email: vsingh@tamu.edu

More information about this series at http://www.springer.com/series/6689

Martina Zeleňáková · Gabriela Hudáková · Agnieszka Stec

Rainwater Infiltration in Urban Areas

Springer

Martina Zeleňáková
Department of Environmental Engineering
Faculty of Civil Engineering
Institute of Environmental Engineering
Technical University of Košice
Košice, Slovakia

Gabriela Hudáková
Department of Environmental Engineering
Faculty of Civil Engineering
Institute of Environmental Engineering
Technical University of Košice
Košice, Slovakia

Agnieszka Stec
Department of Infrastructure and Water
Management, Faculty of Civil
and Environmental Engineering
and Architecture
Rzeszów University of Technology
Rzeszów, Poland

ISSN 0921-092X ISSN 1872-4663 (electronic)
Water Science and Technology Library
ISBN 978-3-030-34697-3 ISBN 978-3-030-34698-0 (eBook)
https://doi.org/10.1007/978-3-030-34698-0

This Springer imprint is published by the registered company Springer Nature Switzerland AG
The registered company address is: Gewerbestrasse 11, 6330 Cham, Switzerland

Preface

Urban drainage has become one of the most important components of civic amenities. Ensuring long-term functioning of sewerage networks and wastewater treatment plants is, however, connected with constantly rising costs, and today, it is clear that the modern method of urban drainage cannot for financial reasons be universally resolved. The existing methods of urban drainage continually imperil the status of watercourses as well as water sources. Urban hydrology was created in order to improve methods of managing the runoff of precipitation in towns and should protect them from flooding while also protecting public health or the environment. The essence of a future solution consists in finding an acceptable compromise of an alternative solution for draining rainwater from a territory. The content of this work is a study focused on resolving the percolation of water from surface runoff and the confrontation between a field test, laboratory analysis and numerical analysis. By confronting and subsequently proposing conditions for percolation, documents will be created for making urban drainage better and more efficient. The reason for the origin of the subject work follows from the insufficient information on infiltration systems in Slovak technical standards and likewise the lack of support for the percolation of water from surface runoff. This work points out the approaches, principles and fundamentals of a proposal for percolation. The aim of the work is the distribution of scientific knowledge in the field of research and solutions for the percolation of water from surface runoff, with emphasis placed on the retention capacity of the selected territory and the intensity of precipitation.

This book presents a comprehensive study of the percolation of water from surface runoff with an emphasis on the retention capacity and intensity of precipitation. The content of this work is a study focused on resolving the percolation of water from surface runoff and the confrontation between a field test, laboratory analysis and numerical analysis. By confronting and subsequently proposing conditions for percolation, documents will be created for making urban drainage better and more efficient. The aim of the work is the distribution of scientific knowledge in the field of research and solutions for the percolation of water from surface runoff, with emphasis placed on the retention capacity of the selected territory, determination of filtration coefficient and the intensity of precipitation.

Keywords Precipitation • Surface runoff • Percolation of precipitation
Runoff coefficient • Urban drainage

Košice, Slovakia Martina Zeleňáková
Košice, Slovakia Gabriela Hudáková
Rzeszów, Poland Agnieszka Stec

Acknowledgements

The authors would like to thank the reviewers for their constructive comments, namely: Prof. Ing. Petr Hlavínek, C.Sc., MBA, Professor at the Institute of Municipal Water Management, Faculty of Civil Engineering, University of Technology in Brno, Prof. Dr. Miloslav Šlezingr, Professor at the Institute of Water Structures, Faculty of Civil Engineering, University of Technology in Brno and Prof. dr. hab. inż. Daniel Słyś, Professor at the Department of Infrastructure and Water Management, Faculty of Civil and Environmental Engineering and Architecture and Associate Professor Ladislav Tometz from the Institute of Geoscience, Faculty of Mining, Ecology, Process Control and Geotechnologies, Technical University of Košice. The authors are also thankful to Andrew Billingham and David McLean for English correction.

This work has been supported by the Scientific and Educational Grant Agency of the Ministry of Education of the Slovak Republic under project VEGA 1/0609/20.

The authors would also like to thank the publisher Springer Nature for providing the opportunity for this publication.

Contents

About the Authors

Martina Zeleňáková is Associated Professor in the field of Environmental Engineering at the Institute of Environmental Engineering, Faculty of Civil Engineering at the Technical University of Košice, Slovakia. In the framework of her scientific research activities, she has focused on the solution of water management problems, rainwater management, environmental impacts' assessment and separately on the assessment of environmental risks in river basins in relation to flood events, drought and water pollution. The results of her scientific research work have been published in national and international journals, scientific conference proceedings, proceedings of national and international conferences. She is Author of three educational textbooks and six monographs, and she has cooperated in solving national and international projects, of which in six cases she has been Pncipal Investigator of the project.

Gabriela Hudáková is graduated with master's degree and Ph.D. degree in the field of Environmental Engineering at the Institute of Environmental Engineering, Faculty of Civil Engineering at the Technical University of Košice, Slovakia. Her scientific work as well as her thesis was devoted to rainwater management and design of infiltration facilities. She was participating in project implementation regarding the rainwater management in urban areas, and she published the results of her research in scientific proceedings of national and international conferences.

Agnieszka Stec currently works at the Department of Infrastructure and Water Management, Rzeszów University of Technology, Poland. She does research in Environmental Engineering focusing on rainwater management, rainwater utilization and drainage systems in urban areas. Her most recent publication is "The impact of land use and urbanization on drainage system." She is participating in research as well as professional projects.

Chapter 1
Rainwater Management in Urban Areas

Abstract At present, a new trend is arising in regard to approaches which try to renew the developmental treatment of runoff and the quality of water, with growing interest and acknowledgement of the fact that renewing the natural water balance contributes to a healthier environment and improves living conditions in urban landscapes. In the past, rainwater was considered only as an inconvenience; today, it is considered with increased interest as a resource. Yet, despite progress many uncertainties exist in urban hydrology. Additional research is required into the time–spatial dynamic of urban precipitation, particularly in order to improve the short-term prediction of the amount of precipitation. All problems and challenges are covered by the uncertainty of climate changes, which adds the obligation to ensure that systems of rainwater management are adaptable and resistant to changes. Urban hydrology plays an important role in the resolution of these problems.

At present, a new trend is arising in regard to approaches which try to renew the developmental treatment of runoff and the quality of water, with growing interest and acknowledgement of the fact that renewing the natural water balance contributes to a healthier environment and improves living conditions in urban landscapes. In the past, rainwater was considered only as an inconvenience; today, it is considered with increased interest as a resource. Yet, despite progress many uncertainties exist in urban hydrology. Additional research is required into the time–spatial dynamic of urban precipitation, particularly in order to improve the short-term prediction of the amount of precipitation.

All problems and challenges are covered by the uncertainty of climate changes, which adds the obligation to ensure that systems of rainwater management are adaptable and resistant to changes. Urban hydrology plays an important role in the resolution of these problems.

The science of urban hydrology has evolved with the aim of improving urban water management for public health and hygiene, for flood protection, as well as for the protection of the environment and the life cycle of cities. Management of urban drainage requires several disciplines: engineering, environmental science, public health and sociology. Urban hydrology is not even close to being simple and requires the development of new technologies that point out the technical problems in towns and which

M. Zeleňáková et al., *Rainwater Infiltration in Urban Areas*,
Water Science and Technology Library 89,
https://doi.org/10.1007/978-3-030-34698-0_1

correspond to the needs of urban communities. The ability to measure and model hydrological parameters and processes depends on the ability to measure and predict precipitation on a high level in relation to time and spatial accuracy. This requirement is even more necessary in the case of towns. This has led to many innovations in the measurement of precipitation, such as the use of digital rain gauges and the development of precipitation models for application in simulations of precipitation–drainage processes in towns.

In recent decades, a wide scale of approaches has been developed for alleviating hydrological influences as well as impacts on the quality of water as a consequence of urbanization. However, a significant debate still remains over the best approaches, suggesting that the drainage of precipitation and storm water in towns remains a complex and complicated area. A new trend has arisen in regard to more integrated approaches which deal with changes in the regimen of flows and at the same time the quality of water and which attempt to consider rainwater as a source or resource and not only as a problem that has to be eliminated. The development of integrated models for prediction and assessment of the effectiveness of alternative approaches toward drainage of rainwater in cities, considered as part of a broader urban water cycle, has only recently become a major point of interest.

The work tackles an issue from urban hydrology and from the field of handling rainwater. It clarifies the progress achieved in recent years in the field of urban drainage and current trends and deficiencies which represent obstacles for a fully integrated approach to the resolution of urban drainage, so that the needs of both cities and the environment are met. The work is focused on the study of rainwater percolation in selected territories. The achieved results support the development of urban hydrology and resolution of the drainage of rainwater from urbanized territories.

Topicality of the Subject

Management of precipitation water is an important activity not only in agriculture and water management, but also in the construction field. The percolation of precipitation water is a current and important issue, especially in the present in association with climatic change, when it is necessary to slow the runoff of water from land, support the percolation of water into soil and increase the retention capability of the ground. The importance of the presence of water in soil and rock underlines the fact that all the processes in them are closely linked with water, and this is, together with air, nutrients and heat, a primary condition of soil fertility.

An engineer comes across groundwater with a variety of intentions: to utilize it or to protect against it. Therefore, to determine its presence and to conduct quantitative and qualitative analyses is of great importance. The cooperation of hydrologists with geologists, chemists and pedagogues and often with biologists and bacteriologists is, however, necessary. In places where the movement of groundwater takes place under very complex conditions, the laws governing such movements are very complex. The resolution of problems in the conditions of Slovakia requires deeper study of groundwater movement.

Purpose and Aims of the Work

The purpose of the work is the proposal of conditions for percolation, which are derived from an analysis of precipitation, a geological survey in the selected localities, and from existing known equations for calculating the percolation of precipitation from surface runoff, for the needs of urban drainage.

The work is processed in the sense of Slovak legislation, specifically Act No. 364/2004 Coll. on Waters, in the wording of Act No. 394/2009 Coll., Government Regulation No. 269/2010 Coll. by which the requirements for achieving a good state of water are provisioned, and Regulation of the Slovak Ministry of the Environment No. 397/2003 Coll., by which details on measuring the amount of water supplied to public water mains and amounts of released waters are provisioned, on the method of calculating the amount of released wastewaters and waters from surface runoff and on directive numbers of water usage, and existing foreign standards, such as DWA—A 138—Design, construction and operation of percolation facilities for precipitation water, or Czech standard ČSN 75 9010—Percolation facilities for precipitation waters, are included.

The aim of the work is the distribution of scientific knowledge in the field of research and resolving the percolation of water from surface runoff with an emphasis on the retention capacity of the selected territory and intensity of precipitation. The fulfillment of the aim of the work requires resolving tasks through the following subjects:

- setting the proposed values of the amount of precipitation water from the monitored data;
- description of the range and methods of carrying out a geological survey for the percolation of precipitation water;
- resolution and research of the mutual relation of rainfall intensity and percolation of water from surface runoff with the task of determining the percolation period;
- evaluation of alternative solutions for the percolation of rainwater, proposal of an optimal resolution for the given conditions;
- setting the conditions for the percolation of precipitation water;
- interpretation of the scientific contribution of the work and its use in practice.

Methods Used and Process of the Solution

The methods used come from practical experience as well as from knowledge obtained from the available literature and consultation with professionals dealing with the given issue in practice. Figure 1.1 shows an outline of methodology of the research process.

As follows from the diagram, four methods of research are proposed. A field test uses the real measuring of precipitation with parallel percolation. With laboratory analysis, a grading curve is determined and subsequently the filtration coefficient, and numerical analysis deals with the modeling of percolation. Empirical relations and dimensional analysis belong to numerical analysis. A detailed description of

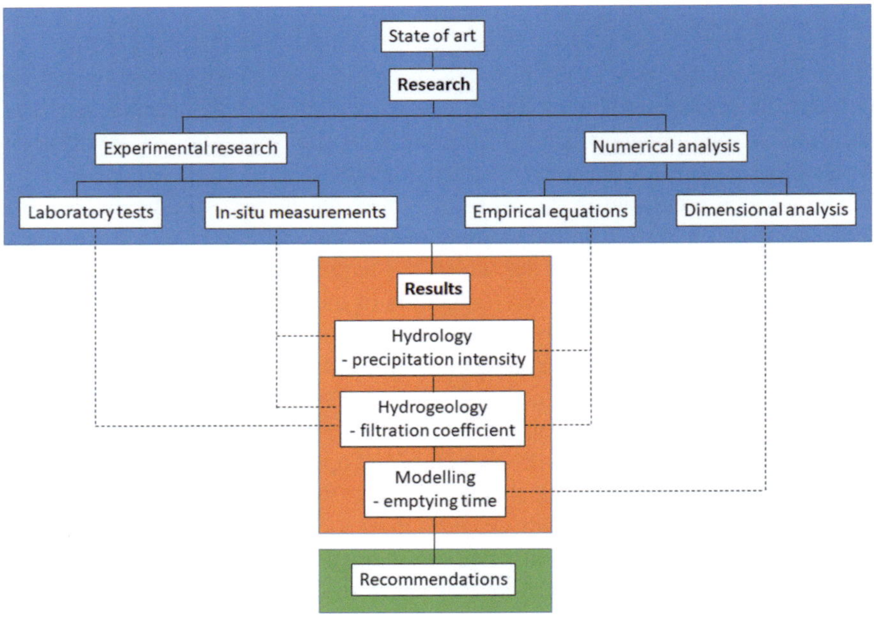

Fig. 1.1 Diagram of the methodological approach of resolving the work

these methods of research is given in Chap. 2. The result of the research is a proposal of conditions for percolation.

Importance of the Work

This work should provide information and resolve the issue of urban drainage. The research problem becomes the subject of a solution, the results of which are documents usable for designers with the proposal of percolation facilities. The importance of this work consists particularly in the proposal of conditions for percolation in the conditions of the Slovak Republic, which supplements the mentioned importance of the work. The conclusions of the work will form the foundations for additional research in the area of urban drainage, since the need for the percolation of precipitation water continues to show a rising tendency.

Synopsis of the Work

The work is composed of four chapters. The introduction contains a brief background summary for the study and outlines the research problem, purpose and aims of the study. In this Chapter, an overview of knowledge from hydrology, an overview of knowledge on the percolation of precipitation from surface runoff in the context of urban drainage and likewise an overview of basic activities from hydropediology and hydrogeology are presented. The Chap. 2 describes a specific study on model territory I—Košice—and model territory II—Prešov—and also their instrumental and software equipping. In this chapter, one can also read about the process of laboratory tests, calculations of the filtration coefficient using empirical relations and dimensional analysis, with the help of which a mathematical–physical model was set up for determining the percolation period. The Chap. 3 describes the research results—a solution for determining the percolation of water from surface runoff. Evaluations of the research results can be read in the Chap. 4. The conclusions of the research and the theoretical and practical contributions of the work, in particular for resolution of the percolation of water from surface runoff, are given in the Chap. 5.

Chapter 2
Urban Hydrology

Abstract Requirements for sustainable long-term environmental protection are rising at present. Not only for financial reasons, but mainly because of concerns about the permanent provision of water resources and the protection of ecosystems, there are doubts about the suitability of the current method of urban drainage. Essentially, such a method of urban drainage is required which ensures not only the protection of humans from the impact of nature but also the protection of the environment from human impacts. The essence of the future solution lies in finding an acceptable compromise between these objectives.

2.1 Management of Urban Hydrology

Hydrology studies the patterns of the occurrence and circulation of water in nature. The causes of the occurrence of hydrological phenomena are monitored and studied using physical and mathematical methods.

Hydrological data serves as a basis for investment activities in almost all sectors of the national economy, especially for water investments such as dams, reservoirs, weirs, water stream regulation or flood protection measures. Detailed hydrological data is required by the agriculture sector, especially in the areas of irrigation, drainage, pond construction and river basin mitigation in general, including soil erosion protection. Investments in water supply for industry and the population and wastewater treatment require hydrological data from groundwater hydrology as well as surface water hydrology. Traffic constructions, especially bridges and cuttings on railways and roads, require very well-determined hydrological data on the average and maximum flow rates in streams. An overview of the main hydrological data relevant to the entire Slovakian economy is contained in the Summary Report on Waters issued annually by SHMI (Dub and Němec 1969).

M. Zeleňáková et al., *Rainwater Infiltration in Urban Areas*,
Water Science and Technology Library 89,
https://doi.org/10.1007/978-3-030-34698-0_2

2.1.1 Precipitation in Urban Hydrology

Research on precipitation in the field of urban hydrology deals with the following topics: precipitation measurements, their prediction and modeling to a specified extent. The study of interactions between land use and precipitation involves the impact of urbanized climate on climatology, which is of primary importance in the context of global urbanization growth (Fletcher et al. 2013).

2.1.2 Precipitation Measurement

Niemczynowicz (1999) considered torrential precipitation as one of the weaknesses in urban hydrology because of the lack of precipitation data needed to determine the time and spatial variations of this natural process. This chapter describes the capabilities of rainfall sensors, meters and radars to accurately measure rainfall according to studies conducted over the past ten years. Lanza and Stagi (2009) tested various precipitators under laboratory conditions with a one-minute interval for different degrees of precipitation to several hundred mm/h and concluded that many precipitation intensifiers met exact WMO specifications (5%), even at one-minute resolution. Ciach (2003) analyzed collision errors based on a sample of data obtained from 15 common precipitators and proposed an analytical expression for standard relative error, depending on the intensity of local precipitation and time span. The representativeness of rainfall data is a key point in determining the spatial variability of rainfall fields at very small scale. Ciach and Krajewski (2006) analyzed data from the EVA-PICONET precipitation monitoring network, which consisted of 25 rain gauges arranged in a regular grid of approximately 3 $\times$ 3 km, increasing spatial correlograms with a short time span: 5 and 15 min. Moreau et al. (2009) studied minor variability with a six-minute time span. The results of this study, however, did not confirm the previous findings: For example, they found a correlation coefficient between 0.5 and 0.6 for ix-minute precipitation versus the correlation coefficient of 0.8 for five-minute precipitation. This difference can be explained by different types of precipitation or by differences in density of precipitation networks. These different results confirm that further research to describe the small variability in precipitation is still needed. Based on the spatial variability characteristics and on the relationship between time and area, Berne et al. (2004) proposed a method that can lead to an estimate of the density of the rain gauge network depending on the concentration and time of precipitation. Urban hydrologists have shown interest in meteorological radars as a means of precipitation mapping with excellent time and spatial resolution. However, conventional radar data is subjected to multiple sources of errors. The identification and correction of these sources of error have become the main focus of research by the scientific community in recent decades. Urban hydrologists contribute to this research effort, as shown, for example: Delrieu and Creutin (1991), Giuli et al. (1991), Einfalt et al. (2004), Upton et al. (2005), Leijnse et al. (2007),

Zinevich et al. (2008), Cummings et al. (2009), Diss et al. (2009), Pedersen et al. (2010), Thorndahl and Rasmussen (2011), Emmanuel et al. (2011, 2012), Cuccoli et al. (2011), Stec et al. (2017).

The rapid development of new technologies predicts a positive future for increasing our ability to capture the time and space variability of rainfall. However, validation of these new techniques will be necessary to ensure that they provide useful data.

In Slovakia, the average annual rainfall varies from less than 500 mm in Galanta, Senec and the eastern part of the Žitný Island to about 2000 mm in the High Tatras (Zbojnícka chata 2130 mm). Relatively low sums of precipitation are in the so-called rain shadows. For this reason, the Spiš basins are relatively dry, protected from the southwest to the northwest by the High and Low Tatras and from the south by the Slovenské Rudohorie range. On average, there will be less than 600 mm of precipitation per year. The amount of precipitation in Slovakia generally falls by approximately 50–60 mm per 100 m of height gain. The northwestern and the northern Slovakia mountains are generally more abundant in atmospheric precipitation than the mountains in the central, southern and eastern regions of Slovakia. This is due to the greater exposure of these mountains to the prevailing northwest flow. Distribution of precipitation during the year is as follows. During the summer season (June–August), it is about 40%, in spring 25%, in autumn 20% and in winter 15% of precipitation (the rainfall is prevalent in summer). The wettest months are June or July, and the least precipitation is in January to March. The great variability of precipitation causes a frequent and sometimes prolonged drought in the lowlands. The Danubian Lowland is one of the smallest areas of Slovakia, on the one hand, with the smallest sums of precipitation (even less than 500 mm per year), but especially because of low rainfall in the summer, and it is also the warmest and relatively hottest area, as a result where there is high potential evaporation. The highest daily precipitation was measured at Salka on Ipeľ during one afternoon when an extraordinarily strong local storm struck on July 12, 1957, dropping 228.5 mm of rain (the total daily amount was even 3.4 mm higher, at 231.9 mm). During the summer period, storms occur in the whole of Slovakia, with large amounts of rainfall (almost every year in Slovakia, there is a daily rainfall of more than 100 mm). The greatest number of days with storms is in the mountains and valleys, where on average there are up to 30–35 such days per year. The least storms occur in the lowlands (SHMI (Slovak Hydrometeorological Institute) 2015).

2.1.3 Impact of Urban Areas on Rainfall

The impact of urbanization on precipitation was identified in the early 1970s (1972) and studied in the METROMEX project, which pointed to reduced summer rainfall in urban areas. This topic has remained an active subject of research. The review by Shepherd (2005) proposed a synthesis of observational and model studies of evoked rainfall changes in cities and presented details of processes that could have produced this impact in cities. Aerosol microphysics and force mechanisms associated with urbanized landscape use play an important role in modifying rainfall events in cities (Ntelekos et al. 2008, 2009). Burian and Shepherd (2005) analyzed data from 19 downpours in Houston's metropolitan area between 1940 and 1999 and found the impact of urbanization. Smith et al. (2005) and Krajewski et al. (2011) used the Hydro-NEXRAD system to generate radar rainfall areas and determine the impact of urbanization on rainfall patterns and spatial heterogeneity. The impact is strongest from midday to midnight during hot periods and causes an increase in precipitation rainfall in urban areas with a windward skew. The climatic impact of urbanized areas on regional storms was confirmed by Niyogi et al. (2010), who rated 91 unique radar-based summer storms, as well as by Rose et al. (2008), who mapped eight years (1995–2003) of average accumulated rainfall using the North American Regional Reanalysis model. Miao et al. (2011), together with Shem and Shepherd (2009), produced model studies confirming evidence of the significant impact of urbanized areas on rainfall variability.

2.2 Urban Hydrology Processes

As a traditional way of transporting water in the drainage of urbanized areas, we consider single or separate sewer systems with gravitational water transport. In the traditional method of drainage, emphasis is placed on the simplicity and reliability of the operation (Hlavínek et al. 2001).

The presence of impermeable surfaces and built-up drainage systems is a basic trigger for changes in hydrology, with an increase in maximum runoffs, annual precipitation volume and runoff variability along with a decrease in infiltration—all of which were thoroughly documented back in the 1960s (Leopold 1968). This chapter explores runoff and urbanization mechanisms (surface and subsurface hydrology as well as evapotranspiration) which affect interactions between water balance components.

2.2.1 Runoff Formation

The formation or the manner of runoff can be divided into long and short terms. The runoff is further divided into surface, subsurface and underground. In practice, the runoff is usually divided into two more detailed entities: direct runoff and base runoff. While the direct runoff includes water running down the surface, the base runoff is largely composed of groundwater alone (Bedient and Huber 2002). More is shown in Fig. 2.1.

This is the sum of the surface and subsurface runoff of water in a river basin. This is runoff which responds directly to the precipitation. Land use determines acquisition properties and interception rates. This significantly influences the intensity of the rapid component of runoff, i.e., direct runoff (Daňhelka 2007).

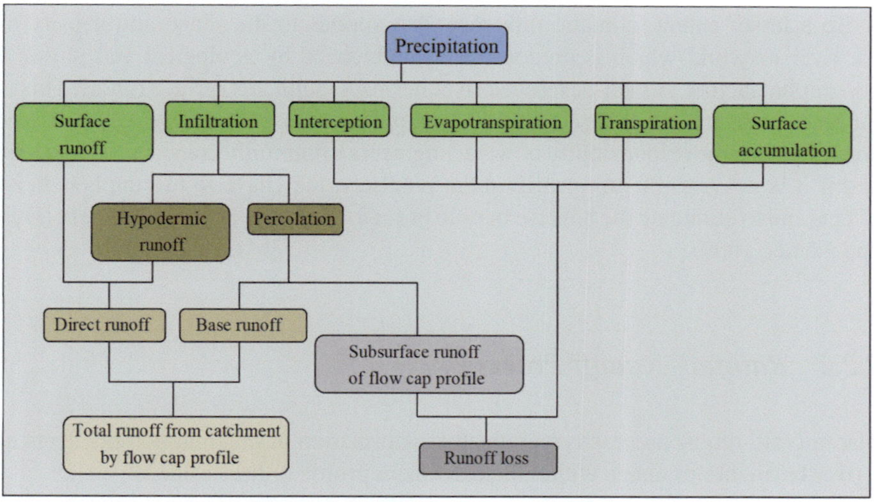

Fig. 2.1 Surface runoff scheme (adopted from ČSN 73 6530 1985)

Some precipitation can soak down and raise the groundwater level, usually a few centimeters. The base runoff in a natural stream is supplied with water from shallow groundwater, and these streams are then part of the hydrographic network. In large natural basins, the basic runoff can be an important component of runoff, and rarely it can be significant in small, urbanized basins where surface runoff prevails (Bedient and Huber 2002). Various factors influence the source that determines the nature of the runoff (Fig. 2.2).

The decisive influence is mainly climatic. In our conditions, the predominant source of surface runoff is atmospheric precipitation. The amount, area and time distribution predetermine the mode of water flows. The precipitation forms part of the cycle of natural water circulation, which ensures the recoverability of surface and groundwater (Słyś et al. 2012).

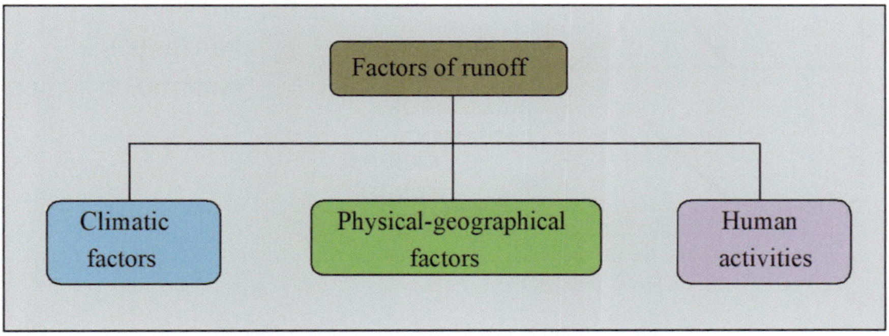

Fig. 2.2 Scheme of drainage factors (adopted from Dub and Němec 1969)

To a lesser extent, climatic influence also applies to the shape and density of the river network, which is predominantly influenced by geological and physical geographic factors. A well-developed river network facilitates surface runoff. This is the reason why it is necessary to artificially supplement insufficiently developed river networks. It is the responsibility of water engineers to carefully consider the impacts, and if it is not possible to quantify them precisely, they have to use the beneficial effects and to eliminate the adverse ones in order to create the best environment (Dub and Němec 1969).

2.2.2 *Rainfall–Runoff Processes*

The rainfall–runoff process is a gradual transformation of precipitation incidents in a river basin into drainage within the catchment profile of the basin.

The rainfall–runoff process itself consists of two partial transformations (Fig. 2.3) (Jandora 2011).

Hydrological transformation is the process, whereby hydrological losses are gradually deducted from rainfall occurring in the basin. These include losses due to evaporation and evapotranspiration (total evaporation from vegetation cover, from plant

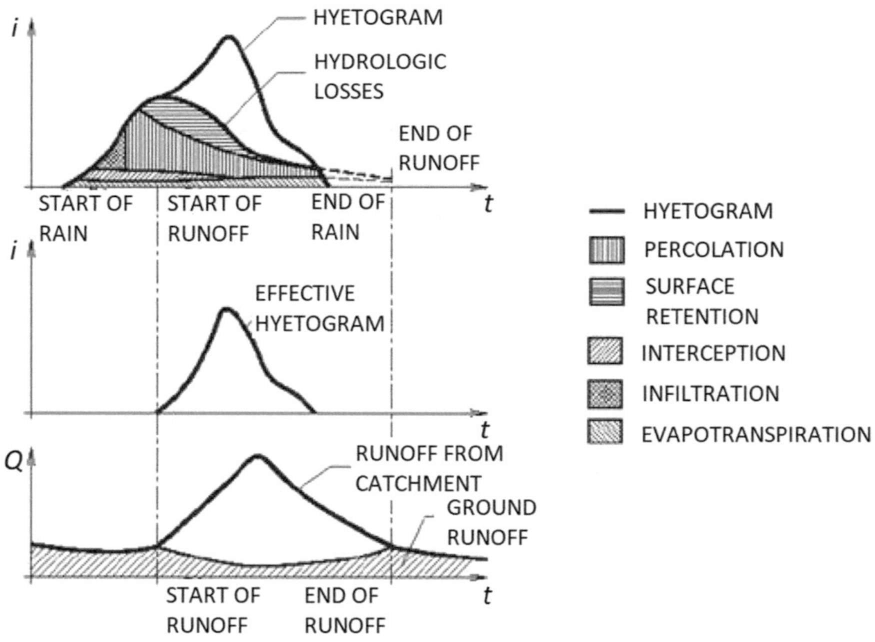

Fig. 2.3 Drainage and drainage scheme in a catchment area (Jandora 2011): *t*—time, *Q*—flow, *i*—intensity of rain

pores and soil); losses due to interception (water retention on the surface of vegetation); losses through moistening; losses through soaking of water into the soil; and losses due to surface retention (surface runoff occurs after uneven terrain is filled with water). Gradual separation of the hydrological losses from the time course of rainfall results in the effective rainfall intensity. The amount of water falling on the surface of the terrain, thus reduced, then flows through the basin as flat surface runoff. This is how the so-called hydraulic transformation proceeds: The flat surface runoff is gradually concentrated in erosive rocks and then in river networks due to drainage within the catchment profile. However, this is not the total runoff which flows through the profile. Part of the total runoff is underground drainage: water that penetrates underground after extreme precipitation. It flows from underground either through the unsaturated zone above the groundwater level or from the saturated zone below the continuous groundwater level in the form of underground drainage into the riverbed. In the unsaturated zone of agricultural cultivated river basins, the soil at tilling depth tends to be very loose and thus has much greater permeability than the soil layer below this horizon. Drainage therefore occurs down the boundary between these layers, and water can flow out of the slope to the surface of the soil. In this case, it is termed hypodermic discharge. The total drainage of water through the basin below the surface of the terrain is called flat underground runoff and is analogous to the surface runoff. Water moves in unsaturated zones down to relatively impermeable subsoil. Sometimes, however, it penetrates through cracks in the impermeable zone to significant depths and can then penetrate to a different catchment area than the original rainfall. Such penetration is called percolation (Jandora 2011). In urbanized basins, the surface runoff is essentially the same as in the natural environment. Its course, however, differs significantly from the precipitation and runoff processes in the natural catchment area. The main causes of these differences are the change in surface permeability and drainage modification caused by urban drainage (Krejčí 2002).

2.2.3 Surface Runoff Processes

Surface water flowing from an area through the river networks comes from rain, snow, groundwater or glaciers. The nature of the water supply source together with the environmental properties of the runoff flow determines the flow mode. Climatic ratios have a decisive influence on the nature of the flow. In the conditions of the Slovak Republic, the predominant source of surface runoff is atmospheric precipitation (Rejdovjanová 2012).

Impermeable surfaces have a tremendous impact on hydrology by eliminating infiltration, thus dramatically increasing the volume of surface runoff. Roads and pavements are generally considered to be impermeable, but in fact their hydrological behavior is based on the intensity and the duration of the precipitation. Ragab et al. (2003a), who studied drainage of rainfall from different types of roofs in a residential area in the UK, found significant differences in drainage and evaporation (evapotranspiration) depending on the slope of the roof, the type of material and its height, which was confirmed by observed variations in the monthly drainage precipitation ratio (that is, from 0.61 to 0.91), depending on the season and the exact characteristics of the roof. Ragab et al. (2003b) then investigated the infiltration and drainage processes at three car parks, on the roadways and on grassland in Wallingford. They came to the conclusion that between 6 and 9% of the rainfall passes through the road surface, and 21–24% of annual precipitation evaporates with a higher percentage during the summer than during the winter. These results were confirmed by Ramier et al. (2011), who studied two different streets in the course of 38 months in Nantes (France). They concluded that losses represented 30–40% of total precipitation and could be caused by evaporation (20% of total precipitation) and soaking (10–20% of total precipitation). Their study was of course dependent on the area and location of the streets, but the results are generally applicable.

The impact of impermeable surfaces on drainage temperature is currently a very original topic of research. Herb et al. (2009) simulated surface drainage and asphalt runoff temperature using climatic data in Minnesota (USA) for six years with 282 individual rainfall events. They found that the temperature of the asphalt runoff was consistent with three parameters: the mean dew point temperature during the storm, the temperature of the air before the storm and the solar radiation before the storm. This conclusion was confirmed by other researchers (Thompson et al. 2008; Janke et al. 2009).

There are also many published findings on the runoff from permeable surfaces in urban areas. The combined effect of removing the top layer of the surface and covering it (due to surface construction, transport and removal of organic matter and vegetation which increases the permeability) means that the permeable surfaces have uncertain precipitation behavior (Shuster and Pappas 2011; Gregory et al. 2006; Booth and Jackson 1997). Subsoil in cities also includes excavations for pipelines and other urban infrastructure which can act as drainage channels and thus affect underground runoff. The effect of land changes due to urbanization depends on geology and on the presence of deep-rooted plants which create macropores in the soil profile (Price 2011). Several authors have concluded that the contribution of permeable areas in urban drainage is relatively small (e.g., Boyd et al. 1993; Ando et al. 1984), while others claim it to be significant (Booth and Jackson 1997; Price et al. 2010).

2.2.4 Subsurface Runoff Processes

Relationships between urbanization and subsurface drainage processes are described in several in-depth studies (Price 2011; Hamel et al. 2013). They appear to be complex, based on the variation of natural drainage characteristics (e.g., geology, topography, vegetation) and the characteristics of urbanization as such (e.g., spatial arrangement of impermeable surfaces and type of drainage) (Hamel et al. 2013). Consequently, subsurface runoff in urbanized basins may increase or decrease. Impervious surfaces result in a decrease in subsurface soaking, but reduction in vegetation in other permeable surfaces can reduce evaporation and potentially increase infiltration. So, the presence of vegetation can restrict the infiltration, thus reducing the subsurface runoff (Price 2011; Hamel et al. 2011). Urban sewerage infrastructure can have important interactions with underground drainage processes. Lowering the groundwater level (Lee et al. 2005) is a common, but not necessary consequence of urban infrastructure such as sewer systems. In the context of the experimental catchment area named Rezé (Berthier et al. 1999), the subsurface runoff represented an average of 14% of the total discharge volume and was most affected by variations in the rainfall rate. These results were also confirmed by Ruban et al. (2007), who observed that the "basic flow" in the sewerage network in the river basin correlated with the level of groundwater. Berthier et al. (2006) proposed a detailed model of interactions between the soil surface in cities and the drainage system in order to simulate case variations in the effluent coefficient. Karpf and Krebs (2011) developed a model to demonstrate groundwater infiltration into sewer systems based on the sewer pipeline classification by year of production. In the research, they created a relationship between these classifications and the take-up behavior. De Silva et al. (2006) created the network exfiltration and infiltration model, which calculated the water level in the sewer pipeline and then applied the Darcy equation to it to determine the exfiltration and infiltration. Interactions between water in the soil and sewer networks were modeled by Goebel et al. (2004) and Endreny and Collins (2009), who investigated the impact of rainwater infiltration on groundwater levels, recommending increased groundwater levels to reduce the risk of contamination. The complexity of subsurface drainage processes means that predicting changes brought on by urbanization and urban infrastructure remains a significant weak point in our capabilities. The development of appropriate tools and a common set of indicators is therefore a very important area of future research, especially in view of the growing importance of the role of subsurface runoff processes in the water environment (Hamel et al. 2013).

2.2.5 *Urban Evapotranspiration*

Evaporation is a major part of the water balance, usually accounting for about 60–95% of average annual precipitation in wooded basins (Zhang et al. 2001). In the urban context, the potential contribution of evaporation is influenced by changes in infiltration, vegetation and urban microclimate. Quantification of urban evaporation is indispensable in determining a city's water balance and has implications for the design of rainfall capture systems and infiltration strategies, along with irrigation management and the urban landscape. Evaporation from urban areas has so far been investigated by climatologists, for the needs of the city's budget in terms of energy management (Coutts et al. 2007). Sailor (2011) provided an overview of methods for estimating anthropogenic heat and humidity in the urban environment. Grimmond et al. (2010) presented an international comparison of 32 ground surface schemes in order to model urban surface processes and energy exchanges, including latent heat exchange. One such model was designed by Grimmond and Oke (2002), who used the Local Scale Urban Meteorological Parameterization Scheme to measure local surface flows from commonly measured data. Coutts et al. (2007) examined changes in local climate and evaporation reduction due to changes in surface processes (and reduction of evaporation) through alterations in land use. Recently, there has been increasing interest in urban evaporation from the hydrological perspective, with the motivation to determine how the urban water balance can be influenced by rainwater management in cities. The study of evaporation in urbanized areas now incorporates soil vegetation atmosphere transfer models, adapted to urban contexts, so that urban evaporation can be identified more precisely, and city budgets can be modified in terms of water and energy economy (Dupont et al. 2006; Noilhan and Malhouf 1996; Lemonsu et al. 2007; Berthier et al. 2006). Wang et al. (2008) developed the urban forest effect hydrology model, which simulates the impact of trees on hydrological conditions at river basin level. Gash et al. (2007) produced encouraging results in estimating the green roof evaporation in the city using a model designed to estimate evaporation from trees. Boegh et al. (2009) also modeled evaporation and runoff on various scales including agricultural, forest and urban use of the landscape. They found that evaporation plays an important role in the country, and also found that in an area with fine vegetation (which is common for urban areas), the parameterization of the vegetation characteristics becomes critical for the precise determination of the evaporation and consequently the runoff estimation. Mitchell et al. (2007) also assessed the impact of urban evaporation on the city budget and confirmed that evaporation is of the utmost importance in estimating the effectiveness of "the most practical" systems in this area. However, this area remains scarcely investigated, with only a few studies quantifying the impact of rainwater management techniques on the total city budget. One of the more recent studies was that by Hamel et al. (2011), who evaluated the relative benefits of infiltration and evaporation by vegetation infiltration systems to assess their impacts on watercourses, groundwater and subsurface runoff. Because of the importance of evaporation in the urban water balance, it is necessary to create new methods for measuring evaporation in cities as well as to continue modeling studies of this process. New techniques such as the use of flow chambers (which can measure the amount of water evaporation as well as other gases such as CO_2) (Rana and Katerji 2000) are starting to be explored. This is undoubtedly an area that requires increased research efforts, regarding not only urban hydrology applications but also the secondary benefits to understanding how urban hydrology can be managed to improve the urban landscape and microclimate (Coutts et al. 2007).

The solution to the problem of evapotranspiration of surface as well as subsurface runoff as part of the rainfall–runoff processes is currently being approached mainly by means of modeling.

2.2.6 *Modeling of Rainfall–Runoff Processes*

Mathematical modeling of rainfall–runoff processes is a very complex problem. Existing models are predominantly aimed at simulating the rainfall–runoff processes in river catchment areas. However, it is not always possible to go into detail in solving this problem. Excessive detail in the approach leads to very complex models which mostly struggle with "data crises" when it is not possible to provide all the required information and a lot of input data tends to be estimated. This reduces the quality and usability of the models. These problems manifest themselves significantly in the modeling of rainfall–runoff processes in extensive river basins. Significantly simplified models, however, provide only very rough estimates of the flow rate in the catchment profile. It is always necessary to look for an acceptable form of simplification (Jandora 2011).

The Department of Water Management at the Faculty of Civil Engineering of the Slovak Technical University in Bratislava offers methodologies and mathematical models for estimating the impact of climate change and changes in land use which may change runoff processes in Slovakian river basins in the coming decades. The updates on offer stem from the growing need to investigate the impact of human activity on hydrological regimes and water quality, particularly in terms of threats from global warming, changes in land use, intensive use of water resources and intensification of agricultural production. The developed methods and model tools offered are based on mathematical, hydrological models simulating runoff processes in river basins. The department's methodology has been developed and tested in many research projects, both nationally and internationally. The modeling results are directly usable in integrated river basin management, water management planning and forecasting in the future, as well as in the design of adaptive measures to mitigate the negative impacts of climate change and land use on water resources (STU 2018).

Mathematical representation of the rainfall–runoff process has a long history, but since the 1980s, thanks to the gradual development of computer technologies, it has become an important tool for hydrologists and water managers, either for operational forecasting or for design purposes (Jeníček 2007). The mathematical model of the rainfall–runoff process represents a simplified quantitative relationship between input and output variables in a particular hydrological system (Daňhelka et al. 2002). It is defined as a system of predominantly physical processes acting on input variables, which are then transformed into an output variable. In mathematical terms, therefore, it is an algorithm for solving the series of equations describing the structure and behavior of the system (Clarke 1973).

The issue of the use of hydrological and one-dimensional hydraulic models is currently dealt with by a number of researchers. Among the Czech authors are Buchtele (2002), Daňhelka et al. (2002), Kulhavý and Kovář (2002), Řičicová and Krejčí (2002) or Starý (2004, 2005). Hydrological and hydraulic models are also used in many specialized workplaces (SHMI, Water Research Institute, SAS—Institute of Hydrology, some university workplaces). The main focus of research and development of all types of hydrological models is to be found abroad, however, particularly in works by Beven (1996, 2001), Bergström (1995), Blöschl and Grayson (2002), Refsgaard and Storm (1996), Smith et al. (2000, 2004) and others who have applied various hydrological models and have significantly contributed to the development of methods describing the rainfall–runoff process. These papers show that emphasis is placed on more accurate processing of rainfall events as the input into the model. Significant space is devoted to the calibration of models, especially to the process of automatically optimizing input parameters. It is also essential to develop methods describing the flow in the unsaturated zone of the soil profile (models of soil moisture). An important contribution to hydrological as well as hydraulic models was the Distributed Model Intercomparison Project, which aimed to compare several rainfall–runoff models. The results were published in the Journal of Hydrology in October 2004 (Smith et al. 2004). The theoretical framework for the whole issue is provided in publications especially by Clarke (1973) and Bear (1972). Many of these and previous approaches are included in WMO methodologies, such as Becker and Serban (1990).

2.3 Integrated Approaches to the Management of Urban Hydrology

Economic growth in developed economies has the effect of disturbing the balance of a whole range of global systems, including water. The past century has revealed the unprecedented possibilities of science and technology, but also the risks resulting from the unrestricted growth of social production and consumption. Rainwater collection systems are commonly used in many countries as sources of water used for gardening, flushing of toilets and even washing, instead of drinking water (Słyś 2009).

2.3.1 Principles and Objectives of Urban Hydrology Management

Urban water management has evolved along with the understanding of its impact on the environment. This new approach has assumed many expressions, including sustainable urban drainage systems (CIRIA (Construction Industry Research and Information Association) 2000), water sensitive urban design (Whelans et al. 1994; Wong 2007) and low impact development (Coffman 2002). These approaches share similar goals:

- to manage the water cycle of the city in a sustainable manner (taking into account surface and groundwater as well as the river basin and impact on soil erosion);
- to keep and return the flow regime as close as possible to the natural character;

- to protect and, as far as possible, restore water quality (surface as well as groundwater);
- to protect and, as far as possible, restore the health of precipitation water;
- to conserve water resources (taking into account the rainwater considered as a resource);
- to strengthen the urban landscape and well-being by incorporating rainwater management measures which offer many benefits for the landscape.

While these approaches increasingly focus on the integration of rainwater management in cities with other parts of the urban water system, hydrology concentrates primarily on the trends and technologies which affect the water cycle of the river basin.

Several authors have attempted to map out conceptual frameworks to protect water from urban effluent. Walsh et al. (2005) suggest:

- reduction of drainage in areas with impermeable surfaces;
- technical solutions to maintain the health of rainwater.

Recently, Burns et al. (2012) reviewed various approaches to rainwater management and agreed on a new approach putting emphasis on protecting and restoring drainage elements which appear to be of ecological importance. This approach to drainage management is based on the identification of hydrological indicators that affect the ecological and geomorphological conditions of incoming water. This approach attempts to maintain the conditions as close as possible to their natural levels (Palhegyi 2010; Tillinghast et al. 2011), respecting the requirement that the urban drainage system must still be able to safely divert peak flow during large storms to minimize the risk to people and protect property from floods. However, identifying indicators which are environmentally relevant and applicable to river basins in different territories remains a challenge.

2.3.2 Technologies for Urban Hydrology Management

Rainwater management technologies are geared to two main activities:

- regulating the quality of water and
- mitigating hydrological changes.

This section focuses primarily on the development of approaches to the management of quantitative aspects of rainwater. Urbanization not only increases the frequency and size of the largest outflows, but also often (though not always) results in a reduction in the amount of groundwater (Price 2011). Appropriate technologies depend on the desired drainage mode:

- rainwater infiltration technologies: including marshes, infiltration pits, tanks, bioreactor systems, porous pathways; the characteristic feature of this group of

technologies is that they help to restore the amount of groundwater through the subsurface runoff;
- rainwater retention technologies: wetlands, ponds, vegetation roofs, rainfall collection (tanks and storage tanks); the characteristic feature of this group is that they help to keep rainwater where it falls, and thus act to reduce runoff.

These techniques can be used either directly or near rainwater sources. Each approach has its advantages and disadvantages: While centralized systems are often effective for the management of the largest outflow (Freni et al. 2010), at the present time it is once again being encouraged to approach drainage in a decentralized way, directly at source, with a proposed holistic approach to recipient recovery (Burns et al. 2012; Shuster et al. 2007).

Basic Infiltration Technologies

Filtration is the soaking of water into the soil. It occurs due to precipitation or to the artificial supply of water to the soil surface. The amount of water infiltrating into the soil per unit of time at time t is referred to as the rate (intensity) of filtration. It is usually given in [mm min^{-1}] or [l s^{-1}]. The total amount of water entering the soil from the start of the infiltration to t is the filtration size [mm] or the amount of filtration [m^3 ha^{-1}]. Surface absorption can occur through these surfaces (Hlavínek 2007):

- grassland;
- damaged gravel surfaces;
- paving blocks;
- permeable paving;
- permeable asphalt.

The variety of natural conditions, as well as technical possibilities, leads to consideration of the optimal way of filtering. The most important is the correct choice of infiltration devices, the most commonly used being the following (Martoň et al. 1991):

- infiltration tanks;
- infiltration channels and trenches;
- sink wells;
- leakage blocks.

The effectiveness of infiltration systems depends to a large extent on local conditions, particularly soil penetration and potential deposition (Le Coustumer et al. 2009; Mikkelsen et al. 1996; Hatt et al. 2009). The clogging of infiltration systems may occur due to:

- incorrectly specified infiltration components (wrong particle size distribution);
- excessive sediment deposition (especially fine sediments) or organic matter which deposits a layer on the surface;
- hydraulic compaction (arrangement and deposition of infiltration components due to water weight) (Gonzalez-Merchan et al. 2010).

Recently, the role of vegetation has been shown to reduce the probability of clogging (through the formation of macropores) (Gonzalez-Merchan et al. 2010; Archer et al. 2002). Hirschman et al. (2008) show a reduction in annual effluent between 40 and 60% in marshes and filter belts containing infiltration components, 50–90% in infiltration ditches and tanks, 40–80% in non-assembled bioreactor systems and 45–75% on porous paths. Filtration-based systems can help restore groundwater as well as subsurface drainage (Endreny and Collins 2009; Shuster et al. 2007). Although there are concerns about groundwater contamination by surface water penetrating via infiltration systems (e.g., Barraud et al. 1999; Pitt et al. 1999), these risks are generally low for most contaminants, although pathogen, salt and heat contamination of groundwater remains unaffected (Weiss et al. 2008; Foulquier et al. 2009). The proposed directives and guidelines, according to Goebel et al. (2004), are needed to ensure that the development of infiltration facilities remains consistent with the sustainable depth of groundwater. There is also a need to have models which are able to predict the impact of infiltration technologies on incoming water and especially on subsurface effluent processes (Hamel et al. 2013).

Basic Retention Technologies

Hydrological modeling technologies can be used to estimate intakes and outflows, thereby evaluating the potential gains of the system in the form of the largest drain reduction. Rainwater capture is an important component of urban hydrology management, but research still needs to be focused on other important challenges, including (Fletcher et al. 2008):

- managing the risks to human health;
- developing an economic framework to quantify all costs and benefits of the life cycle;
- improved, more cost-effective technologies;
- remedying the lack of data in relation to the use of water and energy consumption.

2.3.3 General Empirical Equations for Infiltration

One of the basic tasks in solving hydrological and/or hydrogeological problems is the determination of water infiltration into the soil profile. Soil infiltration capabilities have the characteristics of relief and land management practices, which play a decisive role in formulating criteria for assessing the diversity of different river basins in terms of drainage. Apart from the total amount of water absorbed during infiltration, it is also important to know the time course of the infiltration from the soil surface, the flow of water in the soil profile and the moisture course throughout the profile during the infiltration process. Without detailed knowledge of infiltration, it is not possible to determine the course of effective rainfall in relation to surface runoff (Starý 2005).

Historically, physical aspects of infiltration were neglected, and empirical equations were preferred. This trend is being abandoned today. However, these equations are still popular in the literature and are still practically used (Kutílek and Nielsen 1994).

The equations listed in Table 2.1 simplify the time course of water infiltration into the soil. However, the entire infiltration process is much more complicated and depends on the initial distribution of water moisture in the soil horizon and on the hydraulic properties of the soil.

Infiltration equations are also addressed by other researchers such as Gardner, Widtsoe, Richardson. Kutílek and Krejča (1986) performed the verification of algebraic infiltration equations and came to the following conclusions: The coefficients

Table 2.1 Empirical equation for calculating the infiltration capacity (Starý 2005; Lal and Shukla 2004; ECHO 2018)

Author/s (year)	Equation	Legend
Horton (1940)	$i(t) = i_f + (i_0 - i_f) \cdot e^{-yt}$	$i(t)$—infiltration rate at time t [cm s^{-1}] i_0—initial infiltration capacity [cm s^{-1}] i_f—final infiltration capacity [cm s^{-1}] y—infiltration constant dependent on the type of soil [1 s^{-1}] t—time [s]
Kostjakov (1932)	$i(t) = i_0 \cdot t^{-\alpha}$	α—coefficient dependent on soil properties [−]
Mezencev (1948) Dvořák (1961)	$i(t) = i_0 + (i_1 - i_f) \cdot t^{-b}$	i_1—infiltration capacity $t = 1$ min [cm s^{-1}] b—constant [−]
Holtan (1961)	$i(t) = i_f + cw(IMD - F)^n$	c—factor variable from 0.25–0.8 [−] w—Holtan equation flow factor [−] n—experimental constant = 1.4 [−] F—potential infiltration capacity [−] IMD—initial moisture deficit [−]
Philip (1957)	$i(t) = \frac{1}{2} s_b t^{-0.5} + A_{hv}$	s_b—sorptivity [cm s$^{-0.5}$] A_{hv}—gravity component depending on hydraulic conductivity at saturation [cm s^{-1}]
Dooge (1984)	$i(t) = a(F_{max} - F_t)$	a—constant [−] $F_{\max}$—maximal retention capacity [−] F_t—water quantity retained on soil at time t [−]
Green/Ampt (1911)	$i(t) = k_s\left(1 + \frac{h_0 - h_f}{z_f(t)}\right)$	k_s—hydraulic conductivity at saturation [mm/h] h_0—surface pressure load [mm] h_f—pressure load at the humidity front [mm] z_f—humidity front depths [mm]

in empirical infiltration equations are only parameters of the equalization process, and they are time-unstable and do not have definite physical significance.

Knowledge of hydropedology and hydrogeology is closely related to the study of absorption of water from the surface runoff.

2.4 Basic Knowledge of Hydropedology and Hydrogeology

According to the environment in which the occurrence and movement of water are studied, hydrology can be distinguished as hydrometeorology, dealing with water in the atmosphere; potamology with water in surface water streams; limnology with surface standing waters (in reservoirs and lakes); hydropedology with water in the soil; and geohydrology with water in the geological layers of the Earth's crust. In the latter two cases, focusing on the environment rather than on the occurrence and movement of water is the area of hydropedology and hydrogeology research.

2.4.1 Hydropedology

All the water contained in the soil is called soil water. Soil water fills the soil pores, and at small pore sizes water molecules get into the force fields of the solid phase, so the water properties are affected by the interstitial interface. Soil water has multifaceted significance. In the absence of water, chemical solutions take over and ruin any life in the soil. Soil water contributes to the movement of substances in the soil profile and thus also to soil-type development and the formation of soil properties. It affects the range of physical properties, heat regime and volume changes. The physiological significance of water is clear. Soil water is a basic indispensable factor for soil organisms and plants. It is the medium for receiving nutrients from plants. When assessing physiological aspects, not only the soil water content but also its mobility is important. The soil environment is one of the most important elements of the water cycle in nature. On the one hand, it acts as a storage tank, contributing to the even use of unevenly distributed precipitation. It contributes to runoff retardation, in particular by the conversion of surface water to groundwater. On the other hand, deteriorated surface properties of the soil profile can severely limit drainage from rainfall, and some profile properties can prevent more persistent rainwater percolation and underground water supply (Kutílek 1978).

2.4.2 Hydrogeology

The role and purpose of hydrogeology are to research groundwater resources and evaluate their usable quantities, quality, rational use and protection. Other tasks are performed by hydrogeology in the capture and protection of mineral and thermal waters. They deal with the problems of drainage of deposits in the mining of mineral resources, the occurrence of groundwater in infrastructure and ground constructions, and agricultural activities. Water as a highly mobile element contributes to the movement of pollutants in the atmosphere, on the Earth's surface and below. Many of the current roles of hydrogeology are therefore associated with groundwater pollution and environmental problems. The water management of the Slovak Republic is devoted almost exclusively to water, which falls in the form of precipitation into our territory. Rational water management is therefore of particular importance. According to the Water Law (in Slovakia No. 364/2004 Coll. as amended), groundwater is preferably reserved for the supply of drinking water to the population. Recently, the roles of hydrogeology have been gradually shifting from the search for water resources to their rational use and protection. The resolution of these issues requires good knowledge of geological, hydrological, hydraulic, geophysical and other methods of work, which confirms the position of hydrogeology as one of the borderline sciences between natural and technical disciplines (Fendeková et al. 1995).

2.4.3 Relationship Between Soil and Groundwater

Water is continuously exchanged in liquid or vapor form between the soil profile and water-bearing layers. The intensity of this exchange depends on the factors which determine the groundwater level and groundwater regime. Of these factors, the distribution and the amount of atmospheric precipitation leaking through the soil profile to the groundwater level, the permeability of the soil, the retention capacity of the soil profile, the capillary properties of the soil in the aeration zone, the extent of the groundwater fluctuations and the temperature regime in the soil profile above the groundwater are the most significant. In areas with high precipitation, permeable soils and high groundwater level, the water exchange between the soil profile and the drained horizons represents a significant share of the total water balance of the soil profile. Under conditions of minor precipitation, high retention capacity of the soil and deep groundwater levels, soil and groundwater changes in the soil humidity regime are small or even negligible.

Under natural conditions, three cases of interaction between soil and groundwater can be observed: groundwater control by leaching and discharge of excess groundwater, replenishment of groundwater by penetration of surface water and change of water content in the soil profile caused by groundwater-level fluctuations (Dub and Němec 1969).

2.4.4 Water Movement in the Soil

Water, together with air and other fluids, fills all the cavities of the Earth's crust. The relationship between the rock and the water depends on the size, shape and physical–chemical properties of the surface which contains these cavities. The cavities in the solid rock reach the same dimensions as the surface water and the water flow in them.

The infiltration of surface water beneath the Earth's surface, the movement of water into the soil profile and the heated horizons takes place due to physical forces. The effect of all forces on water in the soil is expressed by the total potential of the soil water, which is the sum of the sorbent, osmotic, capillary (basic), gravitational and possibly also the pressure potential. The soil water potential and soil suction pressure values are experimentally determined at different soil humidities so that a continuous curve of these values can be constructed on the water content (Dub and Němec 1969).

When assessing the subsurface water regime, it is also necessary to know the movement of water, which depends on the permeability k_0 as well as the potential or suction pressure of the soil. The soil permeability is in [cm^2] and depends on the volume, size, coarseness and pore continuity.

When investigating the water movement processes in the soil, it is necessary to know the physical characteristics of the soil as listed in Table 2.2 (Skalová and Štekauerová 2011).

The permeability of loose rock depends predominantly on the size of grains, their distribution, storage density, soil structure and water temperature. The coefficient of filtration k_f (Fig. 2.4) expresses the velocity (its unit is [m s^{-1}]), with which the precipitation water flows through a certain type of environment, in this case the soil. Depending on the type of soil, this velocity can acquire very different (diametrically differing) values ranging from 10^{-1} [m s^{-1}] to 10^{-10} [m s^{-1}]. Within these values, infiltration is said to start if the infiltration velocity of the surrounding soil ranges from 10^{-3} [m s^{-1}] to 10^{-6} [m s^{-1}]. If k_f is greater than 10^{-3} [m s^{-1}], there is a very rapid flow of precipitation water into the groundwater. This results in both the disruption of the temporary storage function of the system and the lack of required soil treatment of precipitation water prior to its entry into groundwater. If, on the other hand, k_f is less than 10^{-6} [m s^{-1}], it is necessary to design a very large storage space, which may be disadvantageous from an economic point of view. In these cases, it is no longer possible to talk about infiltration, but about controlled retention (DWA-A138 2005).

Unless a site hydrogeological survey is carried out, it is recommended to consider that the site contains soil with $k_f = 1.10^{-8}$ [m s^{-1}] (DWA-A138 2005). An important feature in terms of both practical and scientific assessments of the soil is the texture of the soil and its grading composition, because this influences to a large extent the formation and development of the soil and the whole range of soil properties. The decisive extent is affected not only by the physical and chemical properties of the soil, but also by the air, heat and water capacity, structuring ability, permeability to water and air, penetration of plant and trees roots, and orbital resistance. Thus, the

Table 2.2 Physical characteristics of the soil (Skalová and Štekauerová 2011)

Physical characteristics of the soil	Symbol	Unit
Reduced soil mass	ρ_w	[kg/m^3]
Unreduced soil mass	ρ_d	[kg/m^3]
Soil porosity	P	[% vol.]
Drainage soil porosity	P_d	[% vol.]
Soil mass moisture	W	[kg/kg]
Soil volume moisture	Θ	[m/m^3]
Filtration coefficient	k_f	[m s^{-1}]

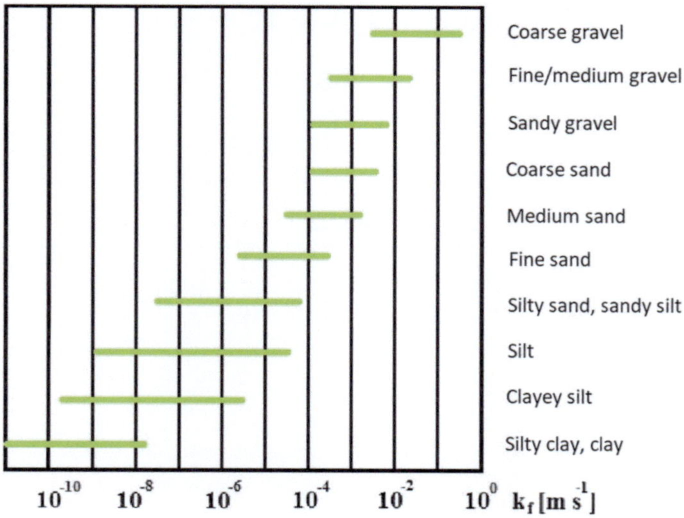

Fig. 2.4 Estimated values of the coefficient of filtration k_f (based on: DWA-A138 2005)

grain composition of the soil, among other characteristics, is very important for any pedological or hydropedological survey.

In order to get an image of the grain composition of the soil, we divide the soil particles into a number of grain-bound intervals, called grain fractions or categories. In Slovakia, detailed division is currently used for the individual fractions listed in Table 2.3.

The grain size results are expressed in a numerical or graphical gradient line, which is the sum of the individual lines. It is possible to determine the percentages of the individual grain fractions and to classify the soil using one of the grading systems, thus assigning a suitable name according to the classification into the relevant group of soil species with a similar texture.

There are several classification ranges and diagrams. The most common in Slovakia is the Kopecký and Novák classification, which is used to solve problems in the soil water regime. The Kopecký classification scale (1899) in Rehák and Jánsky (2000) consists of 18 soil types, which are classified according to the percentage contents of particles of categories I and II compared to categories III and IV. The new classification (1954) is simpler as it only takes into account the content of clay particles, which are decisive for the dynamic processes in the soil, and it divides the soil into seven soil types (Table 2.4) only on the basis of the percentage content of category I (Skalová and Štekauerová 2011).

Table 2.3 Detailed classification of grain fractions (Skalová and Štekauerová 2011)

Grain diameter [mm]	Fraction name	Fine soil categories by Kopecký	
<0.0001 <0.002 0.002–0.01	Colloidal clay Physical clay Very fine silt	Clay particles <0.01	I
0.01–0.05	Silt	Silt	II
0.05–0.1	Powder sand	Powder sand	III
0.1–0.2 0.2–0.5 0.5–2.0	Very fine sand Fine sand Middle sand	Sand	IV
2.0–16.0 16.00–63.00 63.00–125.00 >125.00	Small gravel Middle gravel Rough gravel Stone		

Table 2.4 Soil division—seven soil species (Soil Portal 2018)

Grain category	Particle content <0.1 [mm]	Representation in [%]
Soil light	Sand (0–10%) Loam–sand (10–20%)	6.4
Soil moderate hard	Sand–loam (20–30%) Loam (30–45%)	73.2
Soil hard	Clay–loam (45–60%)	17.1
Soil very hard	Clay (60–75%) Clays (>75%)	3.3

2.4.5 *Water Movement in Rock*

The linear resistance law of the groundwater flow in porous environments was formulated on the basis of the results of experimental research in 1852–1855 by the French hydraulic engineer Henry Darcy.

Darcy's law represents the linear dependence between the infiltration velocity v or the flow rate (Q) and the piezometric gradient i_p, which can be written in the form:

$$Q = k \cdot A \cdot i_p \ [\mathrm{m^3\ s^{-1}}] \tag{2.1}$$

or

$$v = k \cdot i_p \ [\mathrm{m\ s^{-1}}] \tag{2.2}$$

where k is the proportionality constant characterizing the properties of the porous medium and the flow properties of the fluid, and is called the filtration coefficient.

$$k = \frac{K \cdot \rho \cdot g}{\mu} \ [\mathrm{m\ s^{-1}}] \tag{2.3}$$

where K is the coefficient of permeability [m^2], ρ is the density [kg m^{-3}], g is the acceleration due to gravity [m s^{-2}] and μ is the dynamic viscosity [kg m^{-1} s^{-1}].

In technical, particularly constructional practice, several methods are used to determine the filtration coefficient. The choice of method depends on the accuracy of the results to be determined in order to achieve the objective set. The following methods will be described in order to achieve one of the objectives of our research.

2.4.5.1 Laboratory Methods

The value of the coefficient of filtration depends on the characteristics of the soil, especially the diameter d of grains, their shape, the specific surface and its porosity n, and the properties of the water absorbed, especially its bulk density γ and the dynamic viscosity μ, which changes with the water temperature. Among these characteristics, only the grain diameter is most often taken into account. Then, the functional dependency can be written as follows:

$$k = f(d \cdot \gamma \cdot \mu) \tag{2.4}$$

And in dimensional analysis, it can be written in the form:

$$k = C \cdot d^x \cdot \gamma^y \cdot \mu^z \tag{2.5}$$

where C is a non-dimensional constant expressing the influence of the shape of the grains and porosity.

Once the exponents have been solved, the equation takes the form

$$k = C \cdot d^2 \cdot \frac{\gamma}{\mu} \tag{2.6}$$

If the influence of the flowing water is neglected in Eq. (2.6), the value of the coefficient of permeability

$$K = C \cdot d^2 \tag{2.7}$$

by fitting Eq. (2.3) into Eq. (2.7) is then

$$\frac{k \cdot \mu}{\rho \cdot g} = C \cdot d^2 \tag{2.8}$$

and

$$k = \frac{C \cdot d^2 \cdot \rho \cdot g}{\mu} \tag{2.9}$$

This method of determining the filtration coefficient is a quick and inexpensive method which gives only indicative results. It has many forms, and the following formulas can be mentioned by way of example:

Hazen (1893) for materials of different grain shapes by introducing a C_u number (the ratio of the effective diameters d_{60} and d_{10}, which are the diameters representing particle sizes corresponding to 60 and 10% of the weight of the sample (Fig. 2.5).

$$k = \frac{0.01 \cdot d_{10}^2}{C_u} \tag{2.10}$$

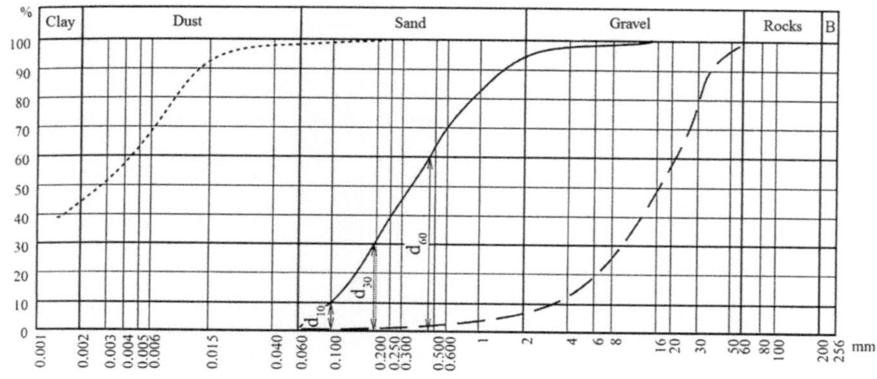

Fig. 2.5 Soil grain curve (Skalová and Štekauerová 2011)

Jáky (1944) for grains in fine-grained soil with the form

$$k = 100 \cdot d_{\text{avrg}}^2 = 100 \cdot d_{50} \tag{2.11}$$

For sandy and gravel alluvial inflows, it is useful to use the formula according to *Bayer* and *Schweiger* (Beyer 1964)

$$k = \frac{\gamma}{\mu} \cdot 1.7 \cdot 103 \left(\frac{d_{60}}{d_{10}} \right)^{-0.2} \cdot d_{10}^2 \tag{2.12}$$

which at 10 °C is $\frac{\gamma}{\mu} \times 7.5 \times 10^{-6} [\text{m s}^{-1}]$.

The practical application of the method in question is demonstrated in Sect. 2.5.2.

2.4.5.2 Hydraulic Methods In Situ

Hydraulic methods are used to determine the basic strict hydraulic parameters of the upper flow coefficient at a known thickness of the drained collector as well as the filtration coefficient. These are field trials, the method of which is to drain water from the well or wells being tested. Pumping and rising examinations are most often used. Where no groundwater is present or cannot be pumped, these tests are replaced with infiltration tests.

Methods of pumping tests are based on the theory of water inflow to a vertical catchment (e.g., well, drill, shaft). The basic formulas for this flow were assembled by Theis (1935), based on the ideal model of radial inflow. To clarify the issue, it is necessary to explain several concepts according to Fig. 2.6.

The static groundwater level is the water level in the well before the pumping of a certain amount of water from the borehole, which we refer to as the yield Q [$\text{dm}^3\ \text{s}^{-1}$, $\text{m}^3\ \text{s}^{-1}$]. The groundwater level in the borehole is reduced to a certain level, the

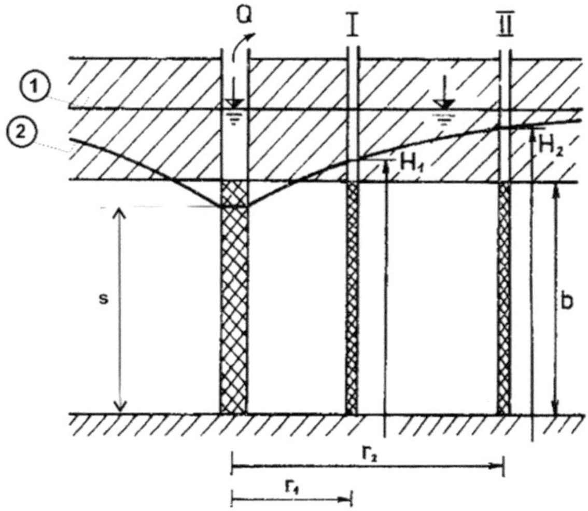

Fig. 2.6 Water supply from the well (Tometz and Burčová 2010): 1—static groundwater level, 2—dynamic groundwater level, *s*—lowering, *Q*—discharge of bore, I—first observation well, II—second observation well, r_1 and r_2—distance of drilling boreholes from well, *b*—the thickness of the layer, H_1 and H_2—piezometric heights

dynamic level (that is, the level created by the hydraulic interference). The difference between the static and the dynamic levels is called the lowering *s* [m].

The share of the yield and the corresponding lowering is characterized as specific (measurable) yield (*q*):

$$q = \frac{Q}{s}[\mathrm{dm^3\ s^{-1},\ m^2\ s^{-1}}] \tag{2.13}$$

where *Q* is the quantity of taken groundwater (flow rate and yield) [$\mathrm{m^3\ s^{-1}}$] and *s* is lowering of the piezometric level at the distance *r* down the axis of the pumped well [m].

When deducing the basic equations of the constant flow of groundwater to the well, Theis (1935) respected ten simplifying assumptions:

- Groundwater flow takes place in an isotropic, homogeneous layer with a given pressure and a perfectly leak-proof insulator.
- The groundwater is not side-bounded.
- The aquifer is flexible and has the same physical properties at all points.
- Water is flexible and has constant properties in time and space.
- The flow is governed by the Darcy filtering law.
- The piezometric level is horizontal prior to sampling (pumping).
- The well is vertical, hydraulically ideal (perfect), without additional resistances, evenly built-in (fitted out), and the inflow is evenly distributed over the vertical.
- The absorbed quantity (yield) is constant over time.
- The radius of the well is near zero (the water volume in the well is negligible).
- Water is released from the rock immediately with a decrease in piezometric tension.

As soon as the water starts being pumped from the well, a pressure drop and piezometric gradient between the well and the drainage layer, which causes the flow of water from the bed into the well, result from the drop in groundwater level in the well. Gradually with the decrease in the groundwater level, a so-called depression cone (space between the original static and dynamic groundwater levels) occurs.

As the piezometric level is reduced, the water from the original volume in the rock expands and is pushed out of the pores due to the elasticity of the drained layer. The water thus released covers the pumped amount, with the pumping length decreasing the piezometric level in the well and its surroundings with the corresponding spread of the depression cone boundary. In the case of the lateral boundary, the decrease in the piezometric level and the spread of the depression cone would continue to infinity in time and space. As a result of radial symmetry, the piezometric level at any point will be only a function of the well distance, time, pumped quantity and hydraulic parameters of the aquifer. The dependence of the decrease in the piezometric level from the time in the described ideal model with the simplifying assumptions is described by Theis's equation:

$$s = \frac{Q}{4 \cdot \pi \cdot T} \cdot W(u) \quad [\mathrm{m}] \tag{2.14}$$

where s is lowering of the piezometric level at distance r down the axis of the pumped well [m], Q is the groundwater quantity [$\mathrm{m}^3\ \mathrm{s}^{-1}$], T is the coefficient of flow [$\mathrm{m}^2\ \mathrm{s}^{-1}$] and $W(u)$ is Theis well function [–].

Hydraulic methods include infusion and injection pressures. Interpretation of the results of infusion and injection tests in drained collectors can theoretically be interpreted as an interpretation of the results of the sampling tests with appropriate sign changes (negative-level decrease depending on negative fluidity). From the time change of the level in the well added to the constant amount of Q, we can theoretically calculate the hydraulic parameters of the test water drain according to the equation:

$$T = 0.183 \frac{Q}{i_s} [\mathrm{m}^2\ \mathrm{s}^{-1}] \tag{2.15}$$

where T is the coefficient of flow [$\mathrm{m}^2\ \mathrm{s}^{-1}$], Q is the extracted yield, volume flow [$\mathrm{m}^3\ \mathrm{s}^{-1}$], i_s is the straight line of directive (slope) [m].

In practice, the whole issue is complicated by undesirable effects, starting with replenishment of water into the borehole from the surface (physicochemical changes in the borehole wall and in the casing zone). These effects stem from changes in the state of the test object producing results which can differ significantly from the natural characteristics of the tested collector. For this reason, drainage and pressure testing of drained collectors may be considered less than appropriate. It is very difficult to interpret the results of infusion and injection tests in undrained rocks, where the greatest influences mainly of capillary forces and lack of saturation are applied, and where neither the actual extent nor shape of the space is affected by the flow of water from the borehole (Jetel 1982).

The evaluation of the inlet test data according to the principles of continuous flow is a positive step in determining the hydraulic parameters of the rock. Necessary evidence for such an assessment is sufficient quality of data, for only then is it possible to achieve tangible results. In hydrogeological practice, the most commonly used method is the so-called direct transformation method.

The direct transformation method uses logarithmic approximation of the Theis well function (Jacob 1944), which is described in detail in textbooks and manuals (see, e.g., Jetel 1982).

The procedure is generally known. After plotting (Fig. 2.7) in the coordinates $x = \log t$, $y = s$ (or $x = \log t$, where t = dimensionless time), after diagnosis of the representative direct section two points are selected with coordinates (s_1, $\log t_1$) and (s_2, $\log t_2$) of this representative section. The directive of the representative direct section (the slope):

$$i = \frac{s_2 - s_1}{\log t_2 - \log t_1} \tag{2.16}$$

is then inserted into the formula for calculating the flow factor:

$$T = 0.183\frac{Q}{i} \tag{2.17}$$

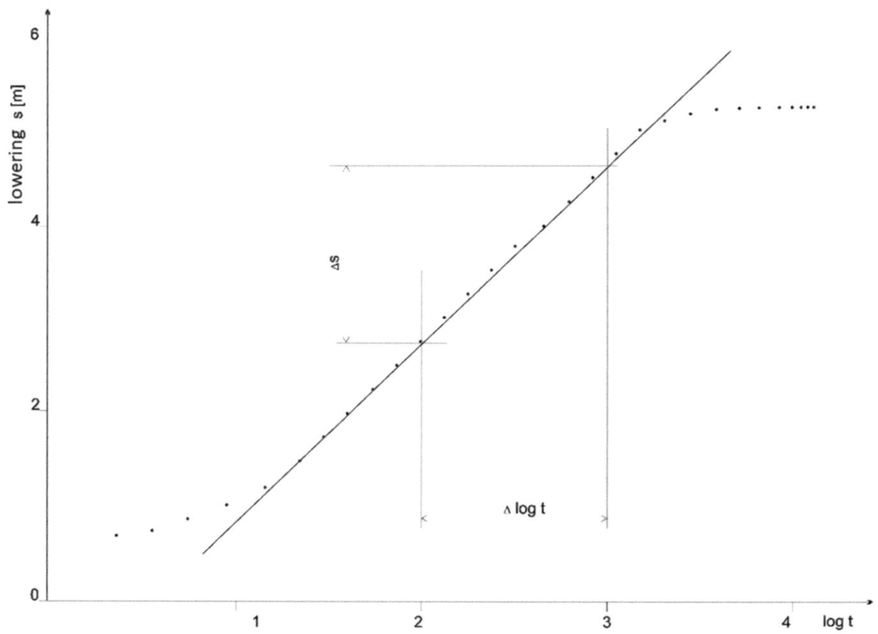

Fig. 2.7 Determination of the line directive from the dependence $s = f(\log t)$

It is evident from the equations that choosing an incorrect direct section with a higher inclination directive as well as the slope of the representative direct section leads to an undervaluation of the coefficient of flow T and the coefficient of filtration and vice versa.

The straight-line transformation method is the basic, simplest method of determining conductive (resistance) hydraulic parameters derived from data corresponding to class I quality. The correctness of the interpretation is, however, conditional upon the correct diagnosis of the representative line section.

In hydrogeological as well as construction practice, there is not always good quality of category I data. In this case, when evaluating the pump test, which provides only category II data derived from strictly hydraulically defined parameters of the rocks (the coefficient of filtration and the coefficient of flow) by approximate (comparative) logarithmic parameters, the index of permeability Z and the Y index must be calculated from the specific yield (Jetel 1968; Jetel and Krásny 1968; Jetel 1974).

The procedure for deriving these parameters along with the procedure for estimating the respective values of the hydraulic parameters is described in detail in the methodological manual on the regional evaluation of the hydrological characteristics of the rocks (Jetel 1985). In the following text, only the basic methodology is given.

The number of wells with category II data, where the values of specific gravity are determined from the archive documentation, is generally much higher than the number of wells with category I data in the investigated area. Considerable yield is the basis for obtaining sufficiently extensive and detailed information on regional permeability and cross-sectional evaluation. The use of approximate parameters derived from the relative expansiveness thus makes it possible to extend significantly also the database used to assess the permeability and flow rate in determining the usable amount of groundwater in the investigated area.

The input data for deriving the approximate logarithmic parameters of permeability and flow rate is the specific yield q, defined as the fraction of the extracted Q—yield and the corresponding lowering in the level s.

$$q = \frac{Q}{s}[\mathrm{dm^3\ s^{-1}}][\mathrm{dm^3 m^{-1}\ s^{-1}}] \tag{2.18}$$

When pumping from a free-flooded aquifer, the yield data is obtained only at $s_n > 1$ m, and the theoretical (extrapolated) standard specific yield q_t is derived as:

$$q_t = q_n \cdot (2M - 1)/(2M - s_n)[\mathrm{dm^3\ m^{-1}\ s^{-1}}] \tag{2.19}$$

where M is the original (by pumping not lowered) thickness of the uninterrupted aquifer [m], s is lowering in free bore [m], $q - Q/s$, the relative yield at $s = s_n \neq 1$ [m].

When lowering the free level s further than $M/10$, the measured reduction must be corrected by Jacob's transformation (Jacob 1944)

$$s_c = s - s^2/2M\ [\mathrm{m}] \tag{2.20}$$

(where s = lowering of the free level measured in the borehole), and in the calculations it is necessary to replace this lowering with corrected lowering s_c. If the thickness of the aquifer M with a free surface is less than 10 m, we also need to apply the Jacob correction for the extrapolated reduction $s = 1$ m in terms of $s > M/10$, so we use the corrected standard metering capacity after calculating the approximate Z and Y parameters and we use the comparison and the combination of the two preconditioning equations expressed by:

$$q_{tc} = q \cdot 2M/(2M - s)\ [\text{m}] \quad (2.21)$$

At the stress level, the dependence of Q on the reduction is theoretically linear, but in practice only a relatively small reduction value s is applied, so for larger lowering the dependence of $Q = f(s)$ on the nonlinear reduction is also changed in the stressing. If we have enough points for the construction of the curve $Q = f(s)$, even in the absence of direct data on the yield, in the case of a decrease $s = 1$ m we can determine the standard metering capacity by extrapolation to $s = 1$ m. If such extrapolation is not possible, we use the parabolic approximation of the curve $Q = f(s)$ in the form of an analogous relationship for the free level, as:

$$q_{tH} = q \cdot (2H_0 - 1)/(2H_0 - s)\ [\text{m}] \quad (2.22)$$

where H_0 is the basal pressure height (the elevation of the static level above the water level) or, in the case of an incomplete bore, elevation of the static level above the lower edge of the tested open section in the well (Jetel 1993). The q and s symbols have the same meaning as in the above equation to extrapolate to q_1 at the free level.

In these equations, we set the yield Q in $m^3\ s^{-1}$ and the lowering s in meters, so that the specific yield is then expressed in $m^2\ s^{-1}$. In practice, we often work with the Q values expressed in $l\ s^{-1}$. If we calculate the quantum yield with Q in $l\ s^{-1}$, it will be expressed in $l/m^{-1}\ s^{-1}$. To distinguish it then in that case, we use the q° symbol.

The approximate logarithmic parameter of permeability derived from the mean yield is the index of permeability Z (Jetel 1968, 1974), defined by:

$$Z = \log(10^9 q/M) = 9 + \log(q/M)\ [-] \quad (2.23)$$

$$Z = \log(10^9 q^0/M) = 6 + \log(q^0/M)\ [-] \quad (2.24)$$

where $q = q_1$ is the standard yield [$m^2\ s^{-1}$], $q^\circ = q_1^\circ$ is the standard yield [l/m^{-1} s^{-1}] and M is the thickness of water level [m].

Where the determination of the thickness of water level M is uncertain or impossible (e.g., in a dissolved rock mass without distinct resolution of the hydrogeological collectors and insulators, or in the case of incomplete bore drilling), instead of the index of permeability Z from the above equations, a replacement value of the openness index of the open section Z_L is defined by:

$$Z_L = \log(10^9 q/L) = \log(10^6 q^0/L)\ [-] \tag{2.25}$$

where L is the length of the tested open section in the collector (or in the studied rock environment) below the static level.

In the case of an incomplete drilling, the thickness of water level M is greater than the active length of the open segment L, so that Z_L is greater than the Z value that would be obtained in the test from the fully opened thickness of M. The theoretical value of the index Z at full opening of the thickness M is denoted by Z_M, and the symbol q_L is used for the value of true real yield at the length of the open section L. The q_M symbol is then marked with the theoretical mean yield of the complete well, so:

$$Z_M = \log(10^9 q_M/M) \wedge Z_L = \log(10^9 q_L/L)\ [-] \tag{2.26}$$

If the actual thickness of M is known, e.g., from a nearby borehole, from a hydrogeologically untested predrill or from a geophysical measurement, it is possible to take advantage of the fact that the Z_M is located within the interval $Z(q_L, M) < Z_M < Z_L$, where:

$$Z(q_L, M) = \log(10^9 q_L/M)\ [-] \tag{2.27}$$

The symbol $Z(q_L, M)$ denotes the fictitious value of the Z index, which would be derived from the assignment of the actual (measured) mean yield of the incomplete bore q_L to the full thickness of the water level M. The unknown Z_M would be assumed in the first approximation using an arithmetic mean when the values q_L, L and M are known:

$$Z_M \cong [Z_L + Z(q_L, M)]/2\ [-] \tag{2.28}$$

At higher M/L ratios, however, this formula overestimates the influence of the inlets from the uncontrolled part of the aquifer. A more appropriate approximation is the weighted average (Jetel 1993):

$$Z_M = [2Z_L + Z(q_L, M)]/3\ [-] \tag{2.29}$$

For very high thickness to open section length ratios (for $M/L > 3$), we recommend using one of two formulas (Jetel 1993). If very good post-mesh communication is expected, and thus still non-negligible inflows from the distant parts of the unopened section, it is recommended to increase the weight of the Z_L value by modifying the equation into the form:

$$Z_M \cong [3Z_L + Z(q_L, M)]/4\ [-] \tag{2.30}$$

If the borehole structure or the expected layer anisotropy indicates the negligibility of the inflows from the distant unopened parts of the aquifer outside the depth section with the received maximum "effective inflow length" $3L$, then the equation for the estimated value $Z = Z_M$ has the form:

$$Z_M \cong [4Z_L - 0.48]/4\ [-] \tag{2.31}$$

to which it is sufficient to simply enter the Z_L value. The derivation of the above formulas is described in Jetel (1993).

A more accurate procedure for calculating the value $Z = Z_M$ from the measured value Z_L or q_L can be found by applying the formula to derive the ratio q_M/q_L (Turcan 1963). A more detailed description of this procedure is given in Jetel (1985, 1993). Another, also more detailed procedure for correcting the yield on the incompleteness of the well, was introduced by Bradbury and Rothschild (1985). Due to the problem of the incompleteness of the borehole in the real rock environment, the correction of the influence of the incompleteness of the borehole by more complicated and more precise methods is not considered necessary in the case of applications of approximate parameters, and for the most part it is sufficient to apply only the above procedures to estimate the index $Z = Z_M$.

In the same way as the defined permeability index Z, the approximate (comparative) flow rate parameter, the Y-flow index, is also derived from the relative yield (Jetel and Krásny 1968; Jetel 1974). It represents a simple logarithmic transformation of the value of the relative yield:

$$Y = \log(10^9 q) = \log q + 9\ [-] \tag{2.32}$$

$$Y = \log(10^6 q^0) = \log q^0 + 6\ [-] \tag{2.33}$$

where q is the standard yield in [$\mathrm{m^2\ s^{-1}}$] and q^0 is the standard yield in [$\mathrm{l\ s^{-1}\ m^{-1}}$].

From comparison of the relationships defining index Z and index Y, the relation follows

$$Z = Y - \log M\ [-] \tag{2.34}$$

(thickness M in meters), but this is valid only for complete boreholes and usable under suitable conditions to determine Z values.

In case of incomplete drilling, the value of the index Y is an indicator of a certain effective flow rate applied at a given stage of opening of the aquifer and does not correspond to the flow rate of the entire water source when the complete thickness is fully opened. The theoretical value of the Y index when opening the entire drainage thickness is indicated by Y_M. Assuming roughly constant permeability across the thickness M, a representative Y_M value from the received $Z = Z_M$ value is $Y_M = Z_M + \log M$.

To facilitate the practical use of approximate logarithmic parameters, Jetel (1985) introduced auxiliary parameters of logarithmic transformation of strictly hydraulically defined coefficients: the coefficient of filtration k and the coefficient of flow

T.

$$Z_k = \log k + 9 \tag{2.35}$$

$$Z_T = \log T + 9 \tag{2.36}$$

where k is the filtration coefficient [m s^{-1}] and T is the coefficient of flow [m^2 s^{-1}].

These transformations place the strictly hydraulic parameters k and T on a common scale with approximate parameters Z and Y, thus allowing their comparison.

Similar meaning in the opposite direction is then given by antilogarithmic transformations of the Z and Y indices

$$k_Z = \text{anti}\log(Z - 9) = 10^{Z-9}\ [\text{m s}^{-1}] \tag{2.37}$$

$$T_Y = \text{anti}\ \log(Y - 9) = 10^{Y-9}\ [\text{m}^2\ \text{s}^{-1}] \tag{2.38}$$

where k is the coefficient of filtration [m s^{-1}], T is the coefficient of flow in [m^2 s^{-1}], Z is the permeability index [−] and Y is the flow index [−].

For the calculation of the flow coefficient from the results of the hydrodynamic tests, the linear functional dependence between the coefficient of flow T and the specific gravity q is applied.

For practical applications, it is appropriate to quantify this relationship in the form of a logarithmic transformation differentiation (Jetel 1985), defined by the relationship

$$d = \log T - \log q\ [-] \tag{2.39}$$

where d is the conversion differential [−], T is the coefficient of flow in [m^2 s^{-1}] and q is the specific yield [l m^{-1} s^{-1}].

By comparing the equations defining the approximate Z and Y parameters with said expression of the logarithmic transformation difference, the form of the basic equations results in the conversion of the values of the approximate parameters to the respective values of the coefficient of filtration k and the coefficient of flow T:

$$T = \text{anti}\log(Y + d - 9) = 10^{(Y+d-9)}[\text{m}^2\ \text{s}^{-1}] \tag{2.40}$$

$$k = \text{anti}\log(Z + d - 9) = 10^{(Z+d-9)}\ [\text{m s}^{-1}] \tag{2.41}$$

where k is the coefficient of filtration [m s^{-1}], T is the coefficient of flow in [m^2 s^{-1}], Z is the permeability index [−], Y is the flow index [−] and d is the transformation differential [−].

Based on the factually analyzed and analytically constructed differences for some data sets in the regional studies performed so far, this empirical equation was used in this study (references of the works where the individual equations were derived are summarized by Jetel (1993):

$$d = 0.07Y - 0.29\ [-] \tag{2.42}$$

where Y is the flow index [−] and d is the transformation differential [−].

This equation was also used for determining the hydraulic parameters T and k for the quaternary river systems of the Košice Basin, both at Košice and Prešov (both study areas used in our research).

Practical application of the methodology is included in Sect. 2.4.

Methods of infiltration (pouring) tests are various and have been applied by several authors in the past in hydropedological surveys, for example Benetín (1958) and Kutílek (1978).

In order to determine the coefficient of filtration of individual types of soils, Nesterov's infiltration test (Benetín 1958) has been judged successful. Two concentrically mounted cylinders made of steel pipes are fitted at one end with a sharp edge, allowing them to be inserted into the test bed of soil. The tubes are usually fitted with probes at the desired depth. Prior to inserting the cylinders into the ground, the probe is cleaned to maintain the intact nature of the test layer. The bottom of the inner cylinder is covered with about 2 cm of a coarse layer of gravel formed by quartz bands with a grain size of 6–8 mm. Both 254- and 358-mm diameter cylinders are filled with water up to 10 cm high. The constant water level in the cylinders is maintained by continuously pouring water from the Marriott container. The duration of the water infiltration in the soil is limited by the settling of the infiltration intensity of the water. For the calculation, the amount of water consumed in the inner cylinder is then used. The depth of water inlet to the ground is determined by drilling using a handheld drill bit (G10) using a 5-cm diameter drilling tool to avoid damaging the collected soil layer.

Determination of tapping and of the hydraulic parameter is then carried out using the formula:

$$k_f = \frac{Q}{A} \cdot \frac{l}{H + l + h_k} \tag{2.43}$$

where k_f is the filtration coefficient [m s^{-1}], Q is the steady water consumption [m^3 s^{-1}], A is the cross-sectional area of inner cylinder [m^2], l is the depth of water infiltration to soil [m], H is the height of water in the inner cylinder and h_k is the adjusted capillary height in the soil [m] (assumed to be about 50% of the maximum capillary height).

Benetín (1958) recommends the following values for individual grain types of soil (Table 2.5).

The issue of blasting tests in wells in hydrogeological research is evaluated in Jetel (1982) and also by Ferris and Knowles (1954) and Verigin et al. (1977).

Table 2.5 Capillary height values for selected soil types

Grain category	Grain with diameter <0.01 mm [%]	Capillary height h_k [m]
Clay–loam soil	45–60	1.00
Sand–loam soil	20–30	0.8
Loam–sandy soil	10–20	0.4–0.6
Sand–clay–fine grain	5–10	0.3
Sand–fine grain	–	0.2
Sand–middle grain	–	0.1
Sand–coarse grain	–	0.05

2.4.6 Subject and Purpose of Dimensional Analysis

For the purposes of the next task of our research, a mathematical–physical model based on application of the Buckingham dimensional analysis method was developed, called the π theorem. The basis of application of the π theorem is to transform the relation that expresses the phenomenon examined in the form of dependence between the physical variables V_1, V_2, ..., V_n, influencing this phenomenon into another, relatively simpler relation between a smaller number of variables and dimensionless arguments π_1, π_2, ..., π_k, compiled from the presented physical variables.

Every complete physical equation expresses the relation between the variables which are relevant to the problem being investigated and always represents the sum of the variables, all of which must have the same dimensions together. This kind of equation is called dimensionally homogeneous.

The complete physical equation can therefore be understood to mean that the equation takes into account all relations between the dimensions of the relevant variables. The requirement for the equation to be dimensionally homogeneous is at the same time a criterion for the correctness of the given equation. It is a necessary but not sufficient criterion, however (Čarnogurská 2000; Čarnogurská 1998; Kožešník 1983).

In general, the complete physical equation which expresses the dependence of the relevant variables V_1, V_2, ..., V_n in different dimensions can be expressed in the form:

$$\varphi[V_1, V_2, \ldots, V_n] = 0 \tag{2.44}$$

From the requirement of dimensional diversity, it follows that the variables V_1, V_2, ..., V_n cannot be inserted in the given equation separately, but in groups having the form:

$$\pi_i = V_1^{X_{11}} V_2^{X_{2l}} \ldots . V_n^{X_{nl}} \tag{2.45}$$

where $i = 1, 2, \ldots, k$.

The variable *xi* is a dimensionless variable called the criterion of similarity or similar number, respectively the invariant of similarity, whose dimension is equal to one, so $[\pi_i] = 1$ (simple number). By introducing dimensionless variables, Eq. (2.44) for the behavior of the system is obtained:

$$\psi[\pi_1, \pi_2, \ldots, \pi_k] = 0 \tag{2.46}$$

where ψ is the symbol of the dependence of the dimensionless arguments on each other.

The dimensionless arguments $\pi_1, \pi_2, \ldots, \pi_k$ must be dimensionally independent; that is, Eq. (2.45) can be written (dimensionally, not functionally) also in the form:

$$[\pi_1]^{u_1} \cdot [\pi_2]^{u_2} \ldots .[\pi_k]^{u_k} = 1 \tag{2.47}$$

where the brackets represent symbolic dimensions.

The last relationship applies if either all $[\pi_1] = 1$ or if $u_1 = u_2 = \cdots = u_k = 0$. Then, the matrix ($k \times n$) is composed of exponents $x_{11}, \ldots, x_{nk}$ by relation (2.45):

$$\begin{Vmatrix} x_{11}\ x_{21}\ x_{31} \ldots x_{n1} \\ x_{12}\ x_{22}\ x_{32} \ldots x_{n2} \\ \vdots \\ x_{1k}\ x_{2k}\ x_{3k} \ldots x_{nk} \end{Vmatrix} \tag{2.48}$$

If its rank $h = k$, the rows of this matrix will be linearly independent, which is also independent of all dimensions (dimensional independence, not functional).

The basis of this dimensional analysis is to find all independent and dimensional arguments π which can be created from n relevant dimensional variables $V_1, V_2, \ldots, V_n$.

The number of relevant dimensional variables $V_1, V_2, \ldots, V_n$ has a total of m dimensionally independent units $Z_1, Z_2, \ldots, Z_m$ where $m < n$.

According to the relationship (2.45) for dimensionless arguments π_i, their dimension is expressed in the form:

$$[\pi_i] = [V_1]^{X_{1i}} x [V_2]^{X_{2i}} \ldots [V_n]^{X_{ni}} = 1 \tag{2.49}$$

In this case:

$$\begin{aligned} [V_1] &= Z_1^{a_{11}}, Z_2^{a_{12}} \ldots Z_m^{a_{1m}} \\ [V_2] &= Z_1^{a_{21}}, Z_2^{a_{22}} \ldots Z_m^{a_{2m}} \\ &\vdots \\ [V_n] &= Z_1^{a_{n1}}, Z_2^{a_{n2}} \ldots Z_m^{a_{nm}} \end{aligned} \tag{2.50}$$

From Eq. (2.50) with equals of the left side $[\pi_i] = 1$ and Eq. (2.50) for unknown $x_{1i}, x_{2i}, \ldots, x_{ni}$, the following conditions are fulfilled:

$$\begin{Vmatrix} a_{11} & a_{21} & a_{31} \ldots a_{n1} \\ a_{12} & a_{22} & a_{32} \ldots a_{n2} \\ \vdots & & \\ a_{1m} & a_{2m} & a_{3m} \ldots a_{nm} \end{Vmatrix} \cdot \begin{Vmatrix} x_1 \\ x_2 \\ \vdots \\ x_n \end{Vmatrix} = 0 \tag{2.51}$$

The abbreviated relationship (2.51) can be written in the form:

$$x = \begin{Vmatrix} x_1 \\ \vdots \\ x_n \end{Vmatrix} 0 = \begin{Vmatrix} 0 \\ \vdots \\ 0 \end{Vmatrix} \tag{2.52}$$

The matrix A of n columns and m lines is a dimension matrix, and its shape is:

$$\begin{matrix} V_1 & V_2 & V_3 & V_n \end{matrix}$$
$$\begin{Vmatrix} a_{11} & a_{21} & a_{31} \ldots a_{n1} \\ a_{12} & a_{22} & a_{32} \ldots a_{n2} \\ \vdots & & \\ a_{1m} & a_{2m} & a_{3m} \ldots a_{nm} \end{Vmatrix} \tag{2.53}$$

Its rank is h. For n unknown $x_1, x_2, \ldots, x_n$, it is possible to obtain only h of independent equations. This is an incomplete system of linear homogeneous equations.

Because the number of unknown variables n is greater than the number of equations, they are not uniquely determined (the system has infinitely many solutions or none). In their determination, the procedure is such that the rectangular dimensional matrix A is divided into two parts. The first one has h columns and h rows (square matrix), with the columns selected so that the matrix determinant is nonzero.

Likewise, vector x is divided into two parts in the following way:

$$\begin{Vmatrix} b_{11} & b_{21} & b_{31} \ldots b_{n1} & b_{h+1,1,\ldots} b_{n,1} \\ b_{12} & b_{22} & b_{32} b_{n2} & b_{h+1,2,\ldots} b_{n,2} \\ \vdots & & & \\ b_{1m} & b_{2m} & b_{3m} \ldots b_{nm} & b_{h+1,h,\ldots} b_{n,h} \end{Vmatrix} \cdot \begin{Vmatrix} y_1 \\ y_2 \\ \vdots \\ y_h \\ y_{h+1} \\ \vdots \\ y_n \end{Vmatrix} = 0 \tag{2.54}$$

Normally, the variables $y_{h+1}, y_{h+2}, \ldots, y_n$ are linearly dependent on the variables y_1, $y_2, \ldots, y_h$. If $(n - h)$ the choice of these variables to be linearly independent together, we obtain $(n - h)$ linearly independent vectors $y \equiv (y_1, y_2, \ldots, y_n)^\tau$ by solving Eq. (2.54). From a practical point of view, it is advisable to select the unknown y_{h+1}, …, y_n to make the following choices:

$$\begin{array}{llll} & y_{h+1} & y_{h+2} & y_n \\ 1.\,\text{option} & 1 & 0\ldots 0 & \\ 2.\,\text{option} & 0 & 1\ldots 0 & \\ \vdots & & & \\ (n-h).\,\text{option} & 0 & 0\ldots 0 & \end{array} \quad (2.55)$$

and the right solution is one which gives small values if possible. The most suitable results are one and zero.

The matrix of independent choices is always square and represents a unit matrix in the expression of the variables. It is always chosen so that its determinant is nonzero $(\Delta \neq 0)$.

Each solution responds according to the relation (2.45) to a dimensionless argument π. In total, it is possible to obtain $(n - h)$ dimensionless arguments, depending on the number of independent choices y_{h+1} to y_n.

In practical tasks, there are often also the relevant variables, the dimensions of which are the same (e.g., a number of lengths or a series of constants characterizing the properties of the substance). If all these variables were included in the basic relationship of the relevant variables, the calculation would become more complex. For this reason, all variables of the same dimension (e.g., $L_1, L_2, \ldots, L_m$) are transformed into dimensional arguments called simplexes:

$$\frac{L_2}{L_1}, \frac{L_3}{L_1}, \ldots \frac{L_m}{L_1} \quad (2.56)$$

These simplexes are simply connected to other dimensionless arguments, complexes. The complexes are compiled from the relevant variables, of which only one length of L_1 is included from all the lengths.

The principle of modeling consists in creating a mathematical–physical model and then deducing the functional dependence on expressed dimensionless arguments, which are generally exponential in nature. Transforming this function into a logarithmic coordinate system produces a linear character that makes it easier to work with the model and makes it easier to determine the linear function parameters.

It follows from the above that at the beginning of the problem-solving it is necessary to find all the relevant variables influencing the process and whose impact on the process is assumed.

References

Ando Y, Musiake K, Takahasi Y (1984) Modelling of hydrologic processes in a small urbanized hillslope basin with comments on the effects of urbanization. J Hydrol 68:61–83

Archer A, Quinton J, Hess T (2002) Below-ground relationships of soil texture, roots and hydraulic conductivity in two-phase mosaic vegetation in South-East Spain. J Arid Environ 52:535–553

Barraud S, Gautier A, Bardin JP, Riou V (1999) The impact of intentional stormwater infiltration on soil and groundwater. Water Sci Technol 39:185–192

Bear J (1972) Dynamics of fluids in porous media. American Elsevier publishing, New York, p 764

Becker A, Serban P (1990) Hydrological models for water—resoures system design and operation. Operational Hydrology Report No. 34. WMO, Geneva, 80 p

Bedient PC, Huber WC (2002) Hydrology and floodplain analysis, 3rd edn. Pretice Hall, London, p 763

Benetín J (1958) Pohyb vody v zemine. Vydavateľstvo Slovenskej akadémie vied Bratislava, pp 85–88

Bergström S (1995) The HBV model. In: Singh VP (ed) Computer models of watershed hydrology. Water Resource Publications, Highland Ranch, pp 443–476

Berthier E, Andrieu H, Rodriguez F (1999) The Rezé urban catchments database. Water Resour Res 35:1915–1919

Berthier E, Dupont S, Mestayer PG, Andrieu H (2006) Comparison of two evapotranspiration schemes on a sub-urban site. J Hydrol 328:635–646

Berne A, Delrieu G, Creutin JD, Obled C (2004) Temporal and spatial resolution of rainfall measurements required for urban hydrology. J Hydrol 299:166–179

Beven KJ (1996) A discussion of distributed hydrological modeling. In Abbott MB, Refsgaard JC (eds) Distributed hydrological modeling. Kluwer, Dordrecht, pp 255–279

Beven KJ (2001) Rainfall-Runoff modelling, the primer. Wiley, Chichester, p 360

Beyer W (1964) Zur Bestimmung der Wasserdurchlässigkeit von Kiesen und Sanden aus der Kornverteilung, vol 14. Wasserwirtschaft-Wassertechnik, pp 165–169

Blöschl G, Grayson R (2002) Flächendetaillierte Niederschlag-Abfluss Modellierung. In: Wiener Mitteilungen – Band 164 – Niederschlag-Abfluss Modellierung-Simulation und Prognose, Technische Universität Wien, pp 33–55

Boegh E, Poulsen RN, Butts M, Abrahamsen P, Dellwik E, Hanse S (2009) Remote sensing based evapotranspiration and runoff modeling of agricultural, forest and urban flux sites in Denmark: from field to macroscale. J Hydrol 377:300–316

Booth DB, Jackson CR (1997) Urbanization of aquatic systems—degradation thresholds, stormwater detention, and the limits of mitigation. J Am Water Resour Assoc 33:1077–1090

Boyd MJ, Bufill MC, Kness RM (1993) Pervious and impervious runoff in urban catchments. Hydrol Sci 38:463–478

Bradbury KR, Rothschild ER (1985) A computerized technique for estimating the hydraulic conductivity of aquifers from specific capacity data. Ground Water 23(2):240–246

Buchtele J (2002) Úvod k metodám a modelům hydrologických předpovědí. In Patera A et al (eds) Povodně: prognózy, vodní toky a krajina. ČVUT, Praha, pp 33–35

Burian SJ, Shepherd JM (2005) Effect of urbanization on the diurnal rainfall pattern in Houston. Hydrol Process 19:1089–1103

Burns M, Fletcher TD, Hatt BE, Ladson A, Walsh CJ (2012) Hydrologic shortcomings of conventional urban stormwater management and opportunities for reform. Landscape Urban Plan 105:230–240

Ciach GJ (2003) Local random errors in tipping-bucket rain gauge measurements. J Ocean Atmos Technol 20:752–759

Ciach GJ, Krajewski WF (2006) Analysis and modeling of spatial correlation structure in small-scale rainfall in Central Oklahoma. Adv Water Resour 29:1450–1463

CIRIA (Construction Industry Research and Information Association) (2000) Sustainable urban drainage systems—design manual for Scotland and Northern Ireland. CIRIA Report No. C521. Dundee, Scotland

Clarke RT (1973) Mathematical models in hydrology. Irrigation and Drainage No. 19, FAO, Rome

Coffman LS (2002) Low impact development: Smart technology for clear water-definitions, issues, roadblocks and next steps. In: Strecker EW, Huber WC (eds) 9th international conference on urban drainage. American Society of Civil Engineers, Portland, Oregon, USA, p 16

Coutts AM, Beringer J, Tapper NJ (2007) Impact of increasing urban density on local climate: spatial and temporal variations in the surface energy balance in Melbourne, Australia. J Appl Meteorol 47:477–493

Cuccoli F, Baldini L, Facheris L, Gori S, Gorgucci E (2011) Tomography applied to radiobase network for real time estimation of the rainfall rate fields. Atmos Res. http://dx.doi.org/101016/jatmosres201106024

Cummings RJ, Upton GJG, Holt AR, Kitchen M (2009) Using microwave links to adjust the radar rainfall field. Adv Water Resour 32:1003–1010

Čarnogurská M (1998) Rozmerová analýza a teória podobnosti a modelovania v praxi. TU, Košice, p 65p

Čarnogurská M (2000) Základy matematického a fyzikálneho modelovania v mechanike tekutín a termodynamike. Vienala, Košice, p 176

ČSN 73 6530 (1985) Názvosloví hydrologie (in Czech). Praha: VÚNM

Daňhelka J, Krejčí J, Šálek M, Špercl P, Zezulák J (2002) Posouzení vhodnosti aplikace srážko-odtokových modelů s ohledem na simulaci povodňových stavů pro lokality na území ČR. ČZÚ, Praha, p 214

Daňhelka J (2007) Operativní hydrologie: hydrologické modely a nejisota předpovědí. Sborník prací ČHMU, Praha, p 104

De Silva D, Burn S, Moglia M, Tjandraamadja G, Gould S, Sadler P (2006) Network exfiltration and infiltration model (NEIMO). In: Wolf L, Morris B, Burn S (eds) Integrating groundwater into water management. IWA, London, pp 34–50 [chapter 2.2]

Delrieu G, Creutin JD (1991) Weather radar and urban hydrology: advantages and limitations of X-band light configuration systems. Atmos Res 27:159–168

Diss S, Testud J, Lavabre J, Ribstein P, Moreau E (2009) Parent du Chatelet J. Ability of a dual polarized X-band radar to estimate rainfall. Adv Water Resour 32:975–985

Dub O, Němec J (1969) Hydrologie. SNTL, Praha, p 380

Dupont S, Mestayer PG, Guilloteau E, Berthier E, Andrieu H (2006) Parameterization of the urban water budget with the submesoscale soil model. J Appl Meteorol Climatol 45:624–648

DWA-A138 (2005) Planung, Bau und Betrieb von Anlagen zur Versickerung von Niederschlagswasser

ECHO (Laboratory of Ecohydrology. Infiltration Process) (2018) [online] [cit. 20.1.018]. http://echo2.epfl.ch/VICAIRE/mod_1a/chapt_5/main.htm

Einfalt T, Arnbjerg-Nielsen K, Golz C, Jensen NE, Quirmbach M, Vaes G (2004) Towards a roadmap for use of radar rainfall data in urban drainage. J Hydrol 299:186–202

Emmanuel I, Andrieu H, Tabary P (2011) Evaluation of the new French operational weather radar product for the field of urban hydrology. Atmos Res 103:20–32

Emmanuel I, Andrieu H, Leblois E, Flahaut B (2012) Temporal and spatial variability of rainfall at urban hydrological scales. J Hydrol 430–431:162–172

Endreny T, Collins V (2009) Implications of bioretention basin spatial arrangements on stormwater recharge and groundwater mounding. Ecol Eng 35:670–677

Fendeková M et al (1995) Základy hydrogeológie. Polygrafické stredisko UK, Bratislava, pp 5–6. ISBN 80-223-0901-X

Ferris JG, Knowles DB (1954) Slug test for estimating permeability. Geolog Surv Wat Note 26, Washington

Fletcher TD, Deletic A, Mitchell V, Hatt BE (2008) Reuse of urban runoff—a review of recent Australian advances and remaining challenges. J Environ Qual 37:116–127

Fletcher TD, Andrieu H, Hamel P (2013) Understanding, management and modelling of urban hydrology and its consequences for receiving waters: a state of the art. Adv Water Resour 51:261–279

Foulquier A, Malard F, Barraud S, Gibert J (2009) Thermal influence of urban groundwater recharge from stormwater infiltration basins. Hydrol Process 23:1701–1713

Freni G, Mannina G, Viviani G (2010) Urban storm-water quality management: centralized versus source control. J Water Resour Plann Manage 136:268–278

Gash JHC, Rosier PTW, Ragab R (2007) A note on estimating urban roof runoff with a forest evaporation model. Hydrol Process 22:1230–1233

Giuli D, Toccafondi A, Biffi Gentili G, Freni A (1991) Tomographic reconstruction of rainfall fields through microwave attenuation measurements. J Appl Meteorol 30:1323–1340

Goebel P, Stubbe H, Weinert M, Zimmermann J, Fach S, Dierke SC (2004) Near natural stormwater management and its effects on the water budget and groundwater surface in urban areas taking account of the hydrogeological conditions. J Hydrol 299:267–283

Gonzalez-Merchan C, Barraud S, Le Coustumer S, Fletcher TD (2010) Suivi de l'évolution de colmatage dans les ouvrages d'infiltration (Monitoring of clogging evolution in the infiltration system). In: Chocat B, Bertrand-Krajewski JL (eds) Novatech. GRAIE, Lyon, France

Gregory JH, Dukes MD, Jones PH, Miller GL (2006) Effect of urban soil compaction on infiltration rate. J Soil Water Conserv 61:117–124

Grimmond CSB, Oke T (2002) Turbulent heat fluxes in urban areas: observations and a local-scale urban meteorological parameterization scheme (LUMPS). J Appl Meteorol 41:792–810

Grimmond CSB, Blackett M, Best MJ, Barlow J, Baik JJ, Belcher SE (2010) The international urban energy balance models comparison project: first results from phase 1. J Appl Meteorol Climatol 49:1268–1292

Hamel P, Fletcher TD, Walsh CJ, Beringer J, Plessis E (2011) Water balance of infiltration systems in relation to their operating environment. In: 12th international conference on urban drainage (ICUD) (Proceedings on CD). Porto Allegre, Brazil

Hamel P, Daly E, Fletcher TD (2013) Source-control stormwater management for mitigating the effects of urbanisation on baseflow. J Hydrol 485:201–213

Hazen A (1893) Some physical properties of sand and grevels. In: 24th annual report, Massachusetts State Board of Healt

Hatt BE, Fletcher TD, Deletic A (2009) Hydrologic and pollutant removal performance of biofiltration systems at the field scale. J Hydrol 365:310–321

Herb WR, Janke B, Mohseni O, Stefan HG (2009) Runoff temperature model for paved surfaces. J Hydrol Eng 1146–55

Hlavínek P et al (2007) Hospodáření s dešťovými vodami v urbanizovaném území. Brno: Ardec s. r. o., 164p. ISBN 80-86020-55-X

Hirschman D, Collins K, Schueler TR (2008) Technical memorandum: the runoff reduction methods. Center for Watershed Protection

Hlavínek P, Mičín J, Prax P (2001) Příručka stokování a čištění, NOEL. ISBN 80-86020-30-4

Jacob CE (1944) Notes on determining permeability by pumping tests under water-table conditions. U.S. Geological Survey Mimeo Report, Washington

Jáky J (1944) Soil mechanics. Egyetemi nyomda, Budapest, Hungaria

Jandora J (2011) Hydraulika a hydrologie - Modul 01 [online], [cit. 25.11.2011]. http://www.unium.cz/materialy/vut/fast/skripta-hydraulika-a-hydrologie-m12253-p1.html

Janke BD, Herb WR, Mohseni O, Stefan HG (2009) Simulation of heat export by rainfall–runoff from a paved surface. J Hydrol 365:195–212

Jeníček M (2007) Rainfall-runoff modeling in small and middle-large catchments—an overview. Geografie-Sborník ČGS 111(3):305–313

Jetel J (1968) A new comparative regional parameter of permeability for hydrogeologic maps. Mém Assoc Int Hydrogéol 8:101–107

Jetel J (1974) Complètement régional de l´information sur les paramètres pétrophysiques en vue de l´élaboration des modèles des systèmes aquifères. Mém Assoc Int Hydrogéol 10(1):199–203

Jetel J (1982) Určování hydraulických parametrů hornin hydrodynamickými zkouškami ve vrtech. Ústřední ústav geologický, Praha, 248 p
Jetel J (1985) Metody regionálního hodnocení hydraulických vlastností hornin. Metod Příruč Ústř Úst Geol 1:47
Jetel J (1993) Stanovenie hydraulických parametrov hornín pre regionálne prognózy. Manuskript, Geofond Bratislava, 177 p
Jetel J, Krásny J (1968) Approximative aquifer characteristics in regional hydrogeological study. Věst Ústř Úst Geol 43(6):459–461
Karpf C, Krebs P (2011) Quantification of groundwater infiltration and surface water inflows in urban sewer networks based on a multiple model approach. Water Res 45:3129–3136
Krajewski WF, Kruger A, Smith JA, Lawrence R, Gunyon C, Goska R (2011) Towards better utilization of NEXRAD data in hydrology: an overview of hydro-NEXRAD. J Hydroinform 13:255–266
Krejčí V et al (2002) Odvodnení urbanizovaných území. Centa spol. s.r.o, Brno. ISBN 80-86020-39-8
Kožešník J (1983) Teorie podobnosti a modelování. ACADEMIA, Praha, p 216p
Kulhavý Z, Kovář P (2002) Využití modelů hydrologické bilance pro malá povodí. VÚMOP, Praha, p 123
Kutílek M, Nielsen DR (1994) Soil hydrology. Catena Verlag, Cremlingen, Destedt
Kutílek M (1978) Vodohopodářska pedologie. ALFA, Praha, p 129p
Kutílek M, Krejča M (1986) Prověření algebraických infiltračních rovnic. Vodohospodářský časopis 6:635–652
Lal R, Shukla MK (2004) Principles of soil physics. Marcel Dekker Inc, New York
Lanza LG, Stagi L (2009) High resolution performance of catching type rain gauges from the laboratory phase of the WMO Field Intercomparison of Rain Intensity Gauges. Atmos Res 94:555–563
Le Coustumer S, Fletcher TD, Deletic A, Barraud S, Lewis JF (2009) Hydraulic performance of biofilter systems for stormwater management: influences of design and operation. J Hydrol 376:16–23
Lee JY, Choi MJ, Kim YY, Lee KK (2005) Evaluation of hydrologic data obtained from a local groundwater monitoring network in a metropolitan city, Korea. Hydrol Process 19:2525–2537
Leijnse H, Uijlenhoet R, Stricker JNM (2007) Hydrometeorological application of a microwave link: 2 Precipitation. Water Resour Res 43:W04417
Leopold LB (1968) Hydrology for urban land planning: a guidebook on the hydrological effects of urban land use. US Geological Survey, Washington DC, 180p
Lemonsu A, Masson V, Berthier E (2007) Improvement of the hydrological component of an urban soil, vegetation, atmosphere, transfer model. Hydrol Process 21:2100–2111
Martoň J et al (1991) Získavanie, úprava, čistene a ochrana vôd – 2. vydanie. Alfa, Bratislava, 648 p. ISBN 80-05-00830-9
Miao S, Chen F, Li O, Fan S (2011) Impacts of urban processes and urbanization on summer precipitation: a case study of heavy rainfall in Beijing on 1 August 2006. J Appl Meteorol Climatol 50:806–826
Mikkelsen PS, Jacobsen P, Fujita S (1996) Infiltration practice for control of urban stormwater. J Hydraul Res 34:827–840
Mitchell VG, Cleugh HA, Grimmond CSB, Xu J (2007) Linking urban water balance and energy balance models to analyse urban design options. Hydrol Process 22:2891–2900
Moreau E, Testud J, Le BE (2009) Rainfall spatial variability observed by X-band weather radar and its implication for the accuracy of rainfall estimates. Adv Water Resour 32:1011–1019
Niemczynowicz J (1999) Urban hydrology and water management: present and future challenges. Urban Water 1:1–14
Niyogi D, Pyle P, Lei M, Arya SP, Kishtawal CM, Shepherd M (2010) Urban modification of thunderstorms: an observational storm climatology and model case study for the Indianapolis urban region. J Appl Meteorol Climatol 50:1129–1144

Noilhan J, Malhouf JF (1996) The ISBA land surface parameterisation scheme. Global Planet Change 13:145–159
Ntelekos AA, Smith JA, Baeck ML, Krajewski WF, Miller AJ, Goska R (2008) Extreme hydrometeorological events and the urban environment: dissecting the 7 July 2004 thunderstorm over the Baltimore MD Metropolitan Region. Water Resour Res 44:W08446
Ntelekos AA, Smith JA, Donner L, Fast JD, Gustafson WI Jr, Chapman EG (2009) The effects of aerosols on intense convective precipitation in the northeastern United States. Quar J R Meteorol Soc 135:1367–1391
Palhegyi GE (2010) Designing storm-water controls to promote sustainable ecosystems: science and application. J Environ Eng 15:504–11
Pedersen L, Einar JN, Madsen M (2010) Calibration of local area weather radar—identifying significant factors affecting the calibration. Atmos Res 97:129–143
Pitt R, Clark S, Field R (1999) Groundwater contamination potential from stormwater infiltration practices. Urban Water 1:217–236
Price K (2011) Effects of watershed topography, soils, land use, and climate on baseflow hydrology in humid regions: a review. Progress Phys Geogr 35:465–492
Price K, Jackson CR, Parker AJ (2010) Variation of surficial soil hydraulic properties across land uses in the southern Blue Ridge Mountains, North Carolina, USA. J Hydrol 383:256–268
Ragab R, Bromley J, Rosier P, Cooper JD, Gash JHC (2003a) Experimental study of water fluxes in a residential area: 1. Rainfall, roof runoff and evaporation: the effect of slope and aspect. Hydrol Process 17:2409–2422
Ragab R, Rosier P, Dixon A, Bromley J, Cooper JD (2003b) Experimental study of water fluxes in a residential area: 2. Road infiltration, runoff and evaporation. Hydrol Process 17:2423–2437
Ramier D, Berthier E, Andrieu H (2011) The hydrological behaviour of urban streets: long-term observations and modelling of runoff losses and rainfall, runoff transformation. Hydrol Process 25:2161–2178
Rana G, Katerji N (2000) Measurement and estimation of actual evapotranspiration in the field under Mediterranean climate: a review. Eur J Agron 13:125–153
Refsgaard JC, Storm B (1996) Construction, calibration and validation of hydrological models. In: Abbot MB, Refsgaard JC (eds) Distributed hydrological modelling
Rehák Š, Jánsky L (2000) Fyzika pôdy. Bratislava, UKBratislava, p 103
Rejdovjanová G (2012). Vsakovanie vôd z povrchového odtoku. In: Seminár doktorandov 2012: zborník príspevkov: 4. ročník: Košice, 09.-10. február 2012. Košice: TU, SvF, p. 146–150
Řičicová P, Krejčí J (2002) Využití hydrologických modelů a perspektivy jejich rozvoje u nás. Počasí – moderní předpovědní metody, prevence a snižování následků katastrof. ČHMÚ, Praha, pp 36–41
Rose L, Stallins JA, Bentley M (2008) Concurrent cloud-to-ground lightning and precipitation enhancement in the Atlanta, Georgia (United States), Urban region. Earth Interact 12:1–30
Ruban V, Rodriguez F, Rosant JM, Larrarte F, Joannis C, Mestayer P (2007) Hydrologic and energetic experimental survey of a small urban watershed. Novatech. GRAIE, Lyon
Sailor DJ (2011) A review of methods for estimating anthropogenic heat and moisture emissions in the urban environment. Int J Climatol 31:189–199
Shem W, Shepherd M (2009) On the impact of urbanization on summertime thunderstorms in Atlanta: two numerical model case studies. Atmos Res 92:172–189
Shepherd JM (2005) A review of current investigations of urban-induced rainfall and recommendations for the future. Earth Interact 9:1–27
SHMI (Slovak Hydrometeorological Institute) (2015) Climate atlas of Slovakia. SHMI, Banská Bystrica, 228 p
Shuster WD, Pappas E (2011) Laboratory simulation of urban runoff and estimation of runoff hydrographs with experimental curve numbers implemented in USEPA SWMM. J Irrigat Drain Eng 137:343–351

Skalová J, Štekauerová V (2011) Pedotransferové funkcie a ich aplikácia pri modelovaní vodného režimu pôd. STU, Bratislava, p 101
Shuster WD, Gehring R, Gerken J (2007) Prospects for enhanced groundwater recharge via infiltration of urban storm water runoff: a case study. J Soil Water Conserv 62:129–137
Słyś D (2009) Potential of rainwater utilization in residential housing in Poland. Water Environ J 23:318–325
Słyś D, Stec A, Zeleňáková M (2012) A LCC analysis of rainwater management variants. Ecol Chem Eng S 19(3):359–372
Smith MB, Koren VI, Wells E, Wang D, Zhang Z (2000) Evaluation of the advantages of the continuous SAC-SMA model over an event API model. In: 15th conference on hydrology 9–14 Jan 2000, Long Beach
Smith MB, Koren VI, Zhang Z, Reed SM, Pan JJ, Moreda F (2004) Runoff response to spatial variability of precipitation: an analysis of observed data. J Hydrol 298:267–286
Smith JA, Baeck ML, Meierdiercks KL, Nelson PA, Miller AJ, Holland EJ (2005) Field studies of the storm event hydrologic response in an urbanizing watershed. Water Resour Res 41:W10413
Soil Portal (2018) Pôdny Portál [online] [cit. 2018-01-20]. http://www.podnemapy.sk/portal/reg_pod_infoservis/pd/pd.aspx
Starý M (2004) Užití umělých neuronových sítí v aplikované hydrologii – zkrácená verze habilitační práce. VÚT, Brno, p 40
Starý M (2005) Hydrologie-Modul 01. VÚT, Brno, p 103
Stec A, Kordana S, Slys D (2017) Analysing the financial efficiency of use of water and energy saving systems in single-family homes. J Clean Prod 151:193–205
STU (Slovenská technická univerzita v Bratislave), Stavebná fakulta. [online] [cit. 2018-01-20]. http://www.svf.stuba.sk
Theis CV (1935) The relation between the lowering of the piezometric surface and the rate and duration of discharge of a well using ground-water storage. Trans Am Geophys Union 2:519–524
Thompson AM, Kim K, Vandermuss AJ (2008) Thermal characteristics of stormwater runoff from asphalt and sod surfaces. J Am Water Resour Assoc 44:1325–1336
Thorndahl S, Rasmussen MR (2011) Marine X-band weather radar data calibration. Atmos Res 103:33–44
Tillinghast ED, Hunt WF, Jennings GD (2011) Stormwater control measure (SCM) design standards to limit stream erosion for Piedmont North Carolina. J Hydrol 411: 185–96
Tometz L, Burčová M (2010) Hydrologeology (in Slovak). TU FBERG, Košice, p 98
Turcan AN Jr (1963) Estimating the specific capacity of a well. U.S. Geological Survey Professional. Paper 450-E. Washington
Upton GJG, Holt AR, Cummings RJ, Rahimi AR, Goddard JWR (2005) Microwave links: the future for urban rainfall measurement? Atmos Res 77:300–312
Verigin NN, Vasiliev SV, Sarkisjan VS, Šeržukov BS (1977) Gidrodinamičeskie i fiziko-chimičeskije svojstva gornych porod. Nerda Moskva
Walsh CJ, Roy AH, Feminella JW, Cottingham PD, Groffman PM, Morgan RP (2005) The urban stream syndrome: current knowledge and the search for a cure. J N Am Benthol Soc 24:706–723
Wang J, Endreny TA, Nowak DJ (2008) Mechanistic simulation of tree effects in an urban water balance model. J Am Water Resour Assoc 44:75–85
Weiss P, Lefevre G, Gulliver JS (2008) Contamination of soil and groundwater due to stormwater infiltration practices. University of Minnesota St. Anthony Falls Laboratory, Minnesota, p 38
Whelans C, Maunsell HG, Thompson P (1994) Planning and management guidelines for water sensitive urban (residential) design. Department of Planning and Urban Development of Western Australia, Perth, Western Australia

Wong THF (2007) Water sensitive urban design; the journey thus far. Aust J Water Resour 110:213–222

Zinevich A, Alpert P, Messer H (2008) Estimation of rainfall fields using commercial microwave communication networks of variable density. Adv Water Resour 31:1470–1480

Zhang L, Dawes WR, Walker GR (2001) Response of mean annual evapotranspiration to vegetation changes at catchment scale. Water Resour Res 37:701–708

Chapter 3
Research on Rainwater Infiltration

Abstract Rainwater management in urban areas takes into account the needs of both the environment and people. The urbanization process has a dramatic impact on water retention hydrology, leading to increased volumes and outflow rates, filtration losses and subsurface flow. The creation of paved surfaces results in drainage of rainwater into a drainage network, which contributes to the reduction of natural filtration. The essence of the solution to urban hydrology lies in finding an acceptable compromise as an alternative solution for the drainage of rainwater from a territory. The subject of the research described in this chapter is a solution to water infiltration from surface runoff and a comparison of the results obtained from field measurements, laboratory analyses and numerical calculations. Such research is necessary mainly due to insufficient information in the technical standards of the Slovak Republic in the field of infiltration systems and rainwater infiltration, as well as solutions to them.

3.1 Model Locations

This part of the book deals with the drainage of rainwater in specific localities based on knowledge and measurement of qualitative and quantitative indicators of rainfall drainage.

The study areas observed in the research are located in two different regions of eastern Slovakia, namely the cities of Košice (Model location I) and Prešov (Model location II) (Fig. 3.1).

Košice is the second largest city in Slovakia and the largest metropolis of eastern Slovakia. Its geographical coordinates are 48°43′12″ N and 21°15′29″ E. The city is situated at an altitude of 208 m a.s.l. From a geological point of view, the entire territory of the Košice Region falls within the zone of the Inner Western Carpathians. The geology of Košice along water streams is formed by quaternary rocks (Holocene) of floodplain sediments and flushings and furthermore quaternary rocks (Pleistocene) of water sprays and loess wedges. There are also lake and river sediments (Geographical Characteristics of the Košice Region 2014).

M. Zeleňáková et al., *Rainwater Infiltration in Urban Areas*,
Water Science and Technology Library 89,
https://doi.org/10.1007/978-3-030-34698-0_3

Fig. 3.1 Study areas: Košice and Prešov (Emergency National Center 2014)

Prešov is the county seat of the Prešov Region and the third largest city in Slovakia. It lies at 49°00′00″ N and 21°14′00″ E, at an altitude of 296 m a.s.l. The subsoil is made of low-lilied soil (skeletal) and glean soil.

3.1.1 Košice—The Technical University Campus

The first area of interest is located in Košice, namely in the area of the Technical University (TUKE) Campus (Fig. 3.2). The research was carried out in the objects marked in red in Fig. 3.2. This is the PK6 object—the Computer Centre of the Faculty of Electrical Engineering and Informatics TUKE and the KIC object—the TUKE Library Information Centre.

At the Computer Centre of the Faculty of Electrical Engineering and Informatics (PK6) (Fig. 3.3), there are infiltration shafts that were made before the research was started. Water from the roof of the building is drained into the infiltration shafts and is subsequently infiltrated into the subsoil. In the TUKE area, about 300 m from the PK6 building, there is a University Library (KIC) building (Fig. 3.3), on the roof of which is a precipitation measuring device (Fig. 3.14)—a rain gauge serving to monitor the amount of precipitation water during the year.

It should be noted that the PK6 project drainage documentation as well as project documentation for the infiltration shafts located at the PK6 building is not available. For this reason, data, such as the technical parameters of the infiltration shafts and the surface of the roof PK6, from which the rainwater is drained into shaft A and shaft B, was determined directly in the terrain.

The total volume of rainfall drained from the roof of the building was found to be discharged by horizontal and vertical pipelines sinking into the infiltration shafts. Due to the location of the pipeline, it was assumed that drainage of rainwater from the roof of the PK6 building into the infiltration shafts is identical. This means that

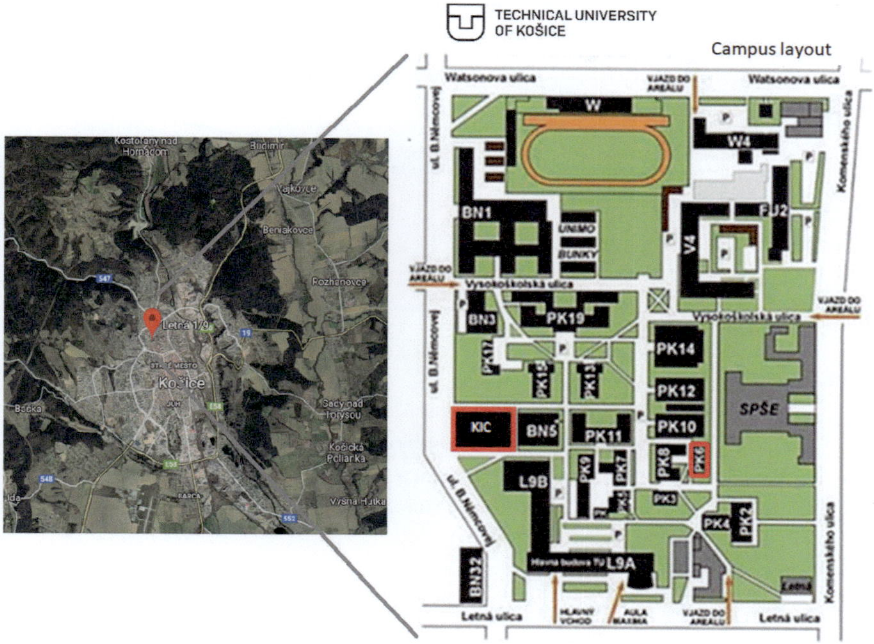

Fig. 3.2 Campus of the Technical University and designation of the sites of the Computer Centre of the Faculty of Electrical Engineering and Informatics (PK6) and the University Library Information Centre (KIC)

Fig. 3.3 Computer Centre of the Faculty of Electrical Engineering and Informatics (PK6) and University Library of the Technical University in Košice (KIC)

the outflow from one half of the roof structure is drained into infiltration shaft A and from the other half into infiltration shaft B (Fig. 3.4).

However, a direct test on the roof of the building refuted the estimate, demonstrating the actual state, which is shown in Fig. 3.5. This means that the outflow from about 1/3 of the roof is drained into infiltration shaft A and approximately 2/3 of the roof area is drained into infiltration shaft B.

The shafts are located on the eastern side of the building. Figure 3.6 shows the

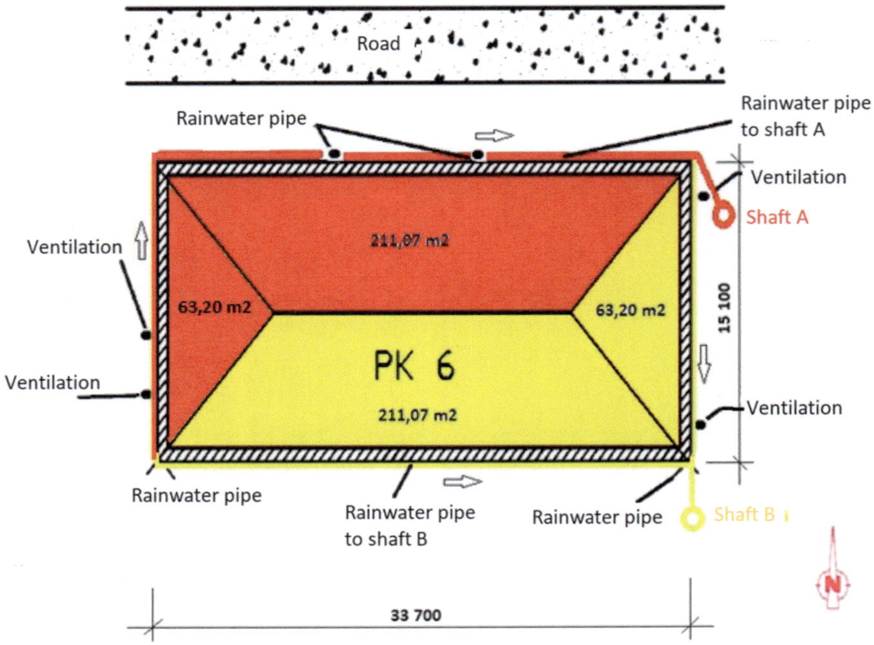

Fig. 3.4 Diagram of the primary estimate of rainwater drainage from the roof of building PK6

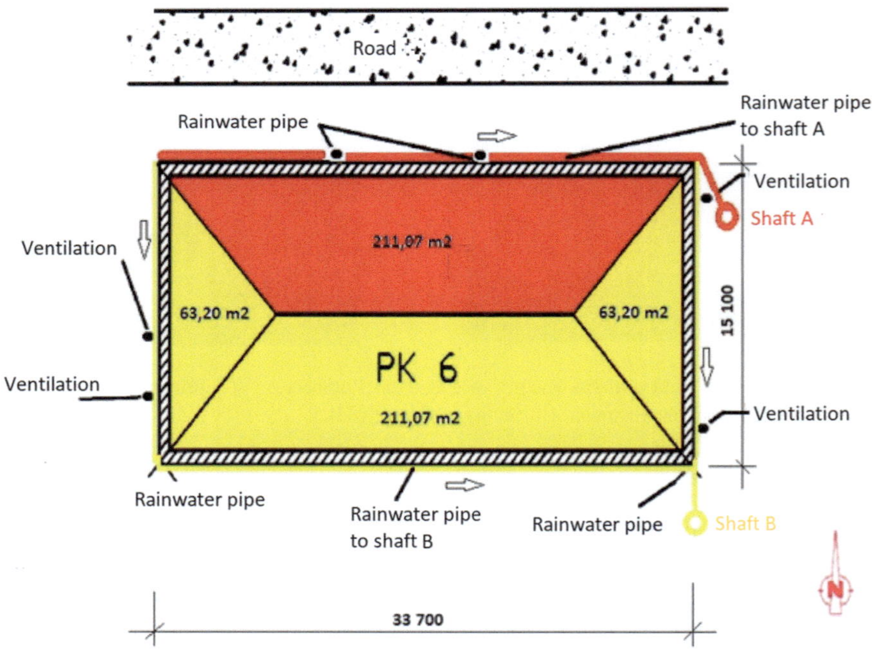

Fig. 3.5 Diagram of the actual drainage of rainwater from the roof of the PK6 object

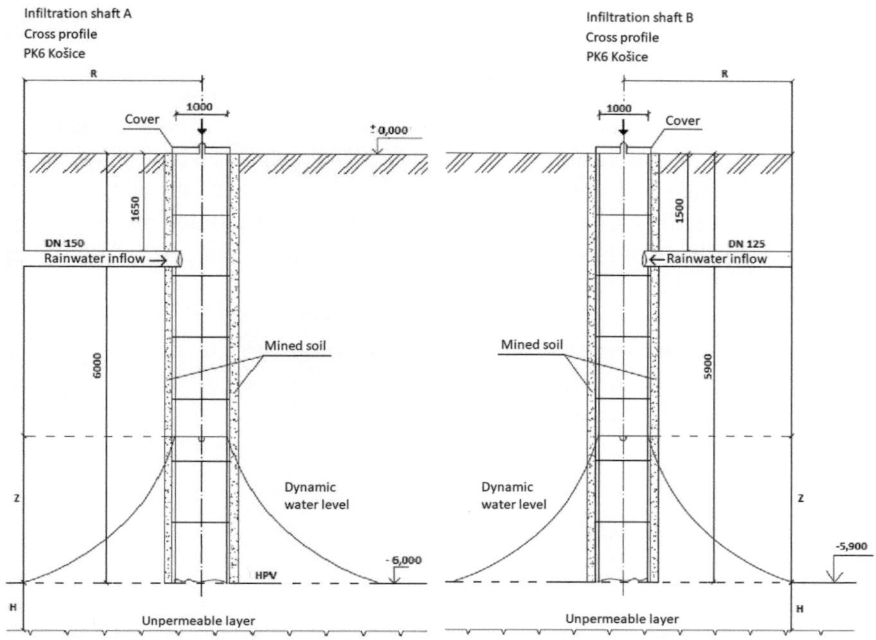

Fig. 3.6 Schematic sections of infiltration shaft A and infiltration shaft B

profiles of infiltration shaft A and infiltration shaft B with the relevant technical parameters (Table 3.1) determined after the field survey.

The shafts are made of concrete cubes with an outside diameter of 1000 mm. Both shafts are secured by a cast iron cover. The technical parameters of shaft A and shaft B are shown in Table 3.1.

The manholes are dimensionally similar, with a difference of more than 100 m^2 of the roof area from which rainwater is drained.

Hydrogeological surveys have been carried out in the past in the area of the Technical University in Košice, suggesting that there are anthropogenic sediments

Table 3.1 Parameters of shaft A and shaft B

Characteristic of the shaft	Shaft A	Shaft B
External diameter	1000 mm	1000 mm
Depth	6000 mm	5900 mm
Pipe dimension (DN) of rainwater supply	DN 150	DN 125
Depth of pipe supply	1650 mm	1500 mm
Roof area drained into the shaft	211.07 m^2	337.47 m^2
Installation of measuring instruments	March 2011	March 2012

in this area—the volumes under which the Hornád river basin is located and below them sediments from the Neogene Period.

The deposits are mostly of gravel, while the walnut component of the natural gravel constitutes the construction waste. The geological wells confirmed the thickness of sediments at 0.5–0.6 m, and a continuous layer of floodplain with a thickness of 4.0–4.5 m was confirmed under the sediments. This is medium to low plasticity clay. Under the flood, sediments are gravels (most often with the addition of fine-grained soil) with a thickness of 5.0–7.0 m. In the quaternary sediments, neogeneous clay gravels and clays with a thickness of 0.7–2.7 m are found.

Within the framework of the project APVV SUSPP-00007-09 "Increasing efficiency of capture and use of precipitation water from surface runoff in order to minimize energy demand", an instrument and software equipment were installed; they are described in the next section (Sect. 2.2).

3.1.2 *Prešov–Šarišské Lúky Location*

The second area of interest for research is located in Prešov, namely in Šarišské Lúky (Fig. 3.7). The reconstruction of a damaged bridge (Fig. 3.8) was carried out in this territory, and a rainwater drainage solution was also provided.

Fig. 3.7 Šarišské Lúky near Prešov

Fig. 3.8 Damaged bridge

The bridge (Fig. 3.9) is located on the I/18 road between Prešov and Kapušany. Communication on the bridge is four-lane and bidirectional. This is a road crossing over the railway and the local Sekčov Road and Route III/6815. The road on the bridge has a one-sided slope of 1.5%. The construction is located outside the village of Ľubotice. Access to the site is possible from existing roads and roads under the bridge (Technical Report 2012).

No engineering geological survey was carried out for the construction site (Technical Report 2012).

The existing bridge is composed of seven profiles consisting of prefabricates of type I 73, a cross section of 115/40, lengths of 30 and 36 m, 15 pcs in cross section. The bridge span is 237.80 m; the total width is 22.20 m, and the width between the

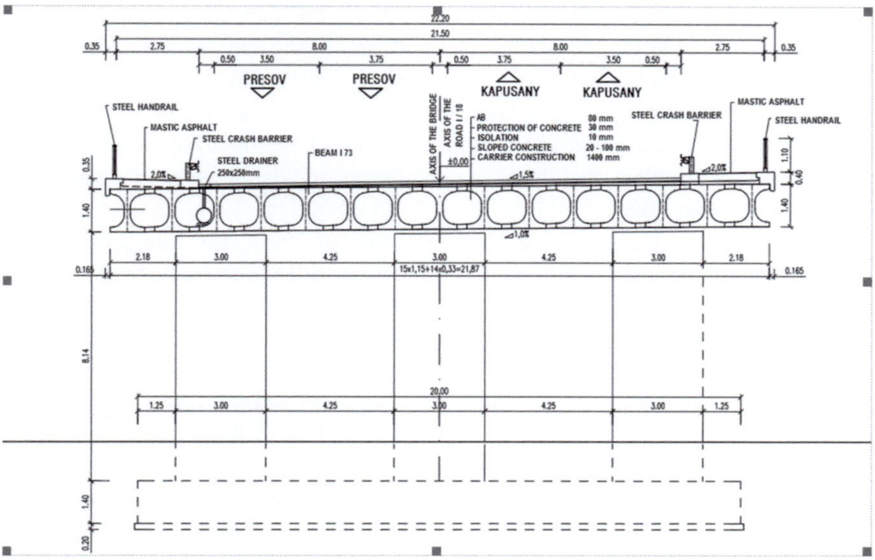

Fig. 3.9 Existing state of the bridge in Šarišské Lúky (Technical Report 2012)

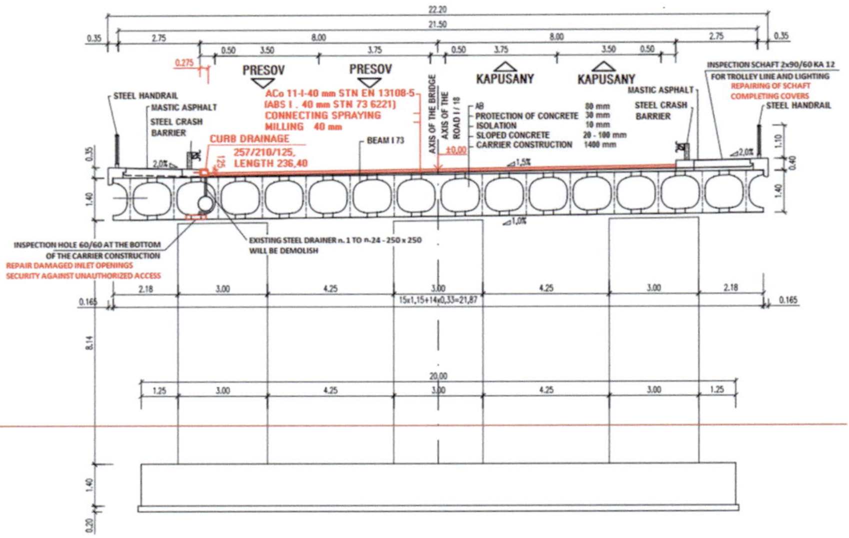

Fig. 3.10 Design of the new state of the bridge in Šarišské Lúky (Technical Report 2012)

bends is 18.0 m. The bridge has a sidewalk on both sides with a deck and railing. In the supporting structure, between the second and third beams in the left, is a drainage pipe to drain water from the drainage system. Leakage of the drainage pipe located in the enclosed space between the beams of the main supporting structure was found at the bottom of the construction. The construction state of the bridge is characterized by the main survey of October 5, 2006, as VI—very bad (Technical report 2012).

The repair of the bridge consisted in the removal of the damaged, non-functional drainage along the entire length of the bridge (Fig. 3.10) and the construction of a new drainage system projecting away from the supporting structure.

The damaged entrance holes in the support structure on the bridge were repaired; they were secured against unauthorized penetration, and hatch covers were added to the shafts in the pavement.

The water on the bridge was drained by means of 250 × 500 mm drains into a collecting pipeline that was fed into a sewer. There were 24 such drainage units on the left side, and two drains were placed on the right at the external supports. The existing drains were destroyed. On the right side, new drains were installed (Fig. 3.11) in the original places, with the result that they were pushed through the supporting structure to the terrain. On the left side, the holes were repaired and the surface drainage of surface water was added via HAURATON ENVIRODECK line drainage channels. There are four lines of curb drainage on the bridge (Fig. 3.12), with a 200 mm diameter drain passing through the support structure, outside the support structure. The main drain is made of polypropylene (PP) and drained through the filter shaft into a collecting gallery made up of HAURATON DRAINFIX BLOC infiltration blocks. The pipeline is embedded in PVC with a diameter of 200 mm. The drainpipe

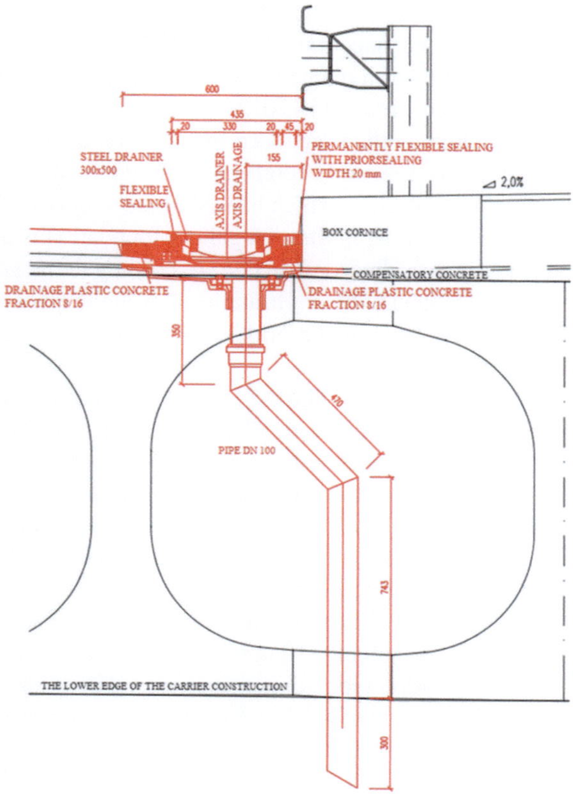

Fig. 3.11 Cross section of the drain (Technical Report 2012)

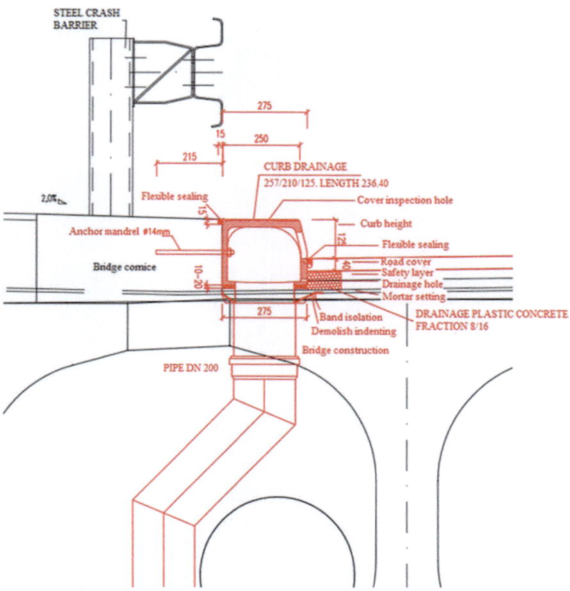

Fig. 3.12 Cross section of the drainpipe (Technical Report 2012)

is lifted to the external supports and in the middle fields to pillars number three and number six. The main drain is made of PP with increased resistance to UV radiation. The drainpipe has a diameter of 200 mm (Technical Report 2012). The infiltration gallery was constructed in 2012.

Infiltration modules are used for filtration and/or controlled drainage of rainwater. They are designed according to the German standard DWA-A138 Design of Percolation Facilities, which is used in the majority of EU countries.

The depth of the infiltration gallery foundation is based on the geological conditions near the bridge construction. According to the geology, the optimal conditions for the placement of the infiltration galley in the layer of sandy clay are approximate. The required layer at the bridge is located at a depth of 2.20–3.10 m (Technical Report 2012).

The HAURATON DRAINFIX BLOC infiltration module (Fig. 3.13) is made of high-strength polypropylene. The spatial structure of the reinforcing elements allows for high strength and load-bearing capacity; compressive strength reaches 600 kN/m^2 in the vertical direction and 120 kN/m^2 in the horizontal direction. Thanks to the achieved values, it is also possible to use infiltration blocks under the communications and parking areas for trucks. Individual blocks can be stored in multiple layers to form a large volume infiltration body even on smaller surfaces. Depending on the proposed traffic load and the friction angle of the surrounding soil, modules can be stored up to a depth of 7.10 m below ground level (Technical Report 2012).

The infiltration block dimensions are 0.60 × 0.80 × 0.33 m (length × width × height). The weight of one block is 11.50 kg. The construction volume of one module is 0.158 m^3, and with a storage capacity of 92%, it is able to capture 0.146 m^3 of

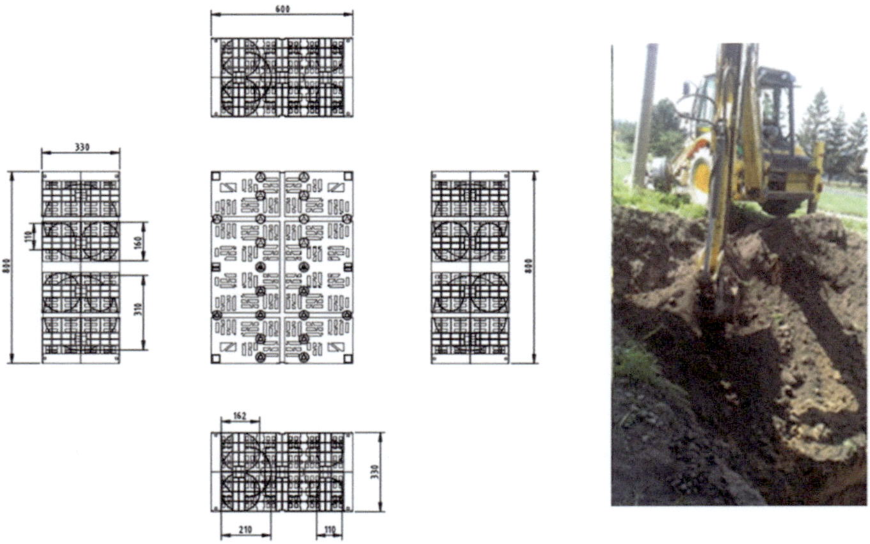

Fig. 3.13 DRAINFIX BLOC infiltration module block and construction

rainwater. The proposed lifespan of the acquisition modules confirmed by long-term testing is 50 years (Technical Report 2012).

The design of the infiltration facility is assessed on the basis of local hydrotechnical and geological parameters for a rain repetition period of 0.2 (1 in 5 years), which ensures high reliability of the intake system (Technical Report 2012).

3.2 Instrument and Software Equipment

3.2.1 Košice—Model Location I

The first measurement of precipitation totals began in March 2011 at the University Library building in parallel with measurement of the amount of rainwater drained from the roof and the amount of rainwater in infiltration shaft A. In March 2012, similar measurements were started in infiltration shaft B.

Rain Gauge MR2H

The MR2H rain gauge is located on the flat roof of the University Library (Fig. 3.14).

The rain gauge has a 200 cm^2 capture area and can detect 0.2 mm of precipitation. It has its own concrete foundation and is centered using a gradient. This rain gauge works on the principle of tilting the boat. When the boat is filled with water, it flips (one tilting = 0.2 mm of precipitation). The scoring device reads these impulses and, based on the conversion factor, indicates the total rainfall. Total precipitation is one of the most common data that is measured using a weather station. The rain gauge may be unheated (they do not measure solid precipitation; therefore, it must

Fig. 3.14 Rain gauge MR2H

Fig. 3.15 Rainlogger

be dismantled or covered before winter operation so that ice does not destroy the interior). The rain gauge located on the roof of the KIC is heated, and therefore, an electrical supply (230 V) must be available to serve for the internal heating of the rain gauge. The heating is switched on by a thermostat. The MR2H operating temperature ranges from −30 to +60 °C.

Rainlogger

At the beginning of the rainfall measurement, data was read manually from a rainlogger (Fig. 3.15) using the Solinst Levelogger software. This is a datalogger that measures the temperature in addition to rainfall. The rainlogger is similar to the other dataloggers (levelogger, barologger) from this manufacturer but varies in the connector for wiring the rain gauge (3-pin connector). The rainlogger reads pulses from the precipitator output and measures the actual amount of rainfall according to the conversion constant.

Its capacity is 40,000 records, which means a 4-month capacity with a 5-min measurement interval. Data is left in the rainlogger until its capacity is reached, or until the data is manually deleted. When the rainlogger switched to automatic measurement, it joined the rain gauge, and the data was transferred to the Fiedler-Magr server, where it is updated daily and can be checked online.

Levelogger and Barologger

At the beginning of the measurement, when the rainfall events were downloaded manually, the water level reading in the A shaft was also manually downloaded using the levelogger measuring device (Fig. 3.16).

The levelogger is a recording device from Solinst, which can record the movement of the water level in groundwater or surface water. The recording unit, the pressure sensor, the temperature sensor and the battery, which lasts up to 10 years, are in one box. The whole device operates on the principle of the so-called Faraday's cage—it is resistant to external electromagnetic interference. All communication (programming, downloading) is done optically. Leveloggers use so-called indirect compensation: Atmospheric pressure compensation is done by software using an

Fig. 3.16 Levelogger

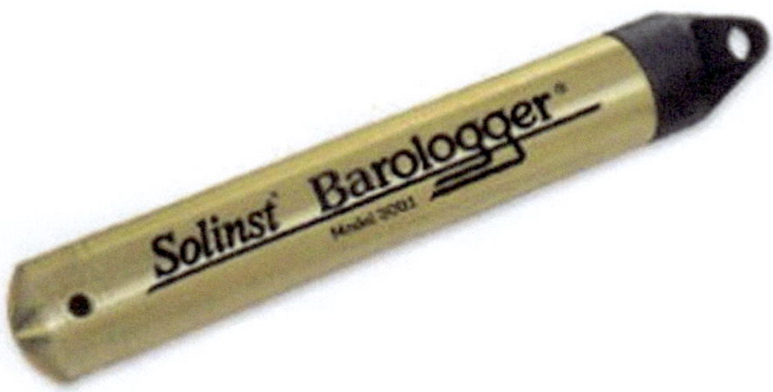

Fig. 3.17 Barologger

external barometric sensor—a barologger (Fig. 3.17). The accuracy of the Levelogger Gold level measuring is 0.05% of the total range (the most accurate range is 2.5 mm). It is able to record data at intervals of 0.5 s to 99 h.

A barologger is a recording device that measures the same parameters as the Levelogger Gold, but the pressure measurement is adapted to atmospheric pressure (not inserted). It is used for the barometric compensation that we perform after downloading the data in the software. The barologger has an algorithm that calculates the atmospheric pressure at the height of the water column. The software then simply subtracts the values measured by the levelogger from the values measured by the barologger, and the result is the actual height of the water column in the shaft. The accuracy of the barologger pressure sensor is 0.002% of the total range.

Data is downloaded from the levelogger and barologger manually using the Solinst Leveloger software. The communication interfaces through which leveloggers can communicate are USB, RS232 and SDI-12.

Solinst Levelogger Software

Datalogger data is downloaded continuously, until the last day of the month. The Solinst Levelogger software must be installed to perform a reading. This simple software makes it easy to work with dataloggers. It has the ability to download data repeatedly. Dataloggers are connected to a notebook using an optical cable. The optical cable is used to transfer information between the server and the computer using an infrared port (Fig. 3.18).

Figure 3.19 shows the Solinst Levelogger software, which shows the input data

Fig. 3.18 Connecting a laptop with dataloggers using an optical cable (Fiedler 2008)

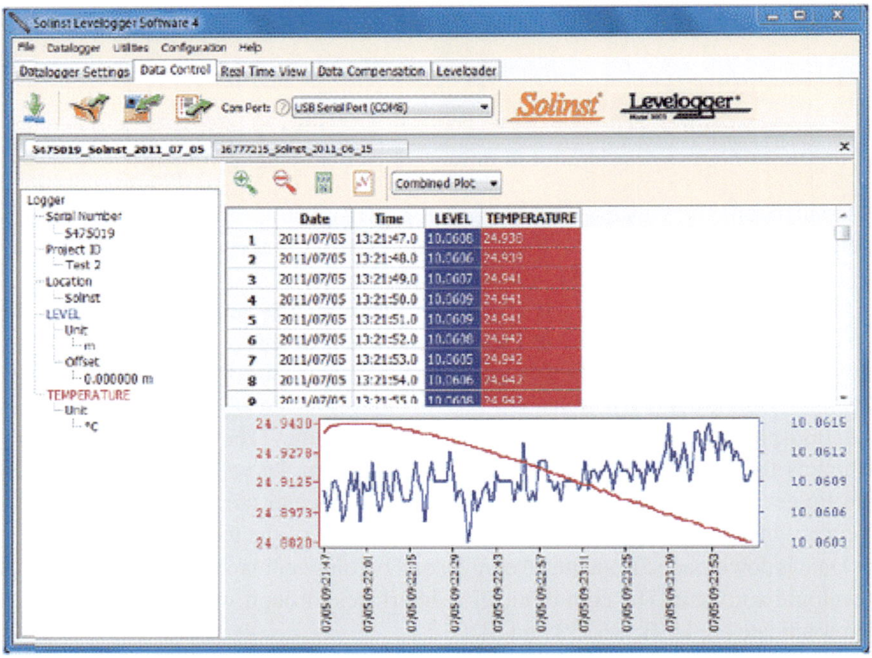

Fig. 3.19 Output from the Solinst Levelogger software

and data about the project on the left, and the measured data in the right side, both in the form of a table and in the form of a graph. Data is downloaded to the computer by clicking on the screen icon or by clicking the levelogger button. The software allows instant viewing of data in a chart or in a spreadsheet format.

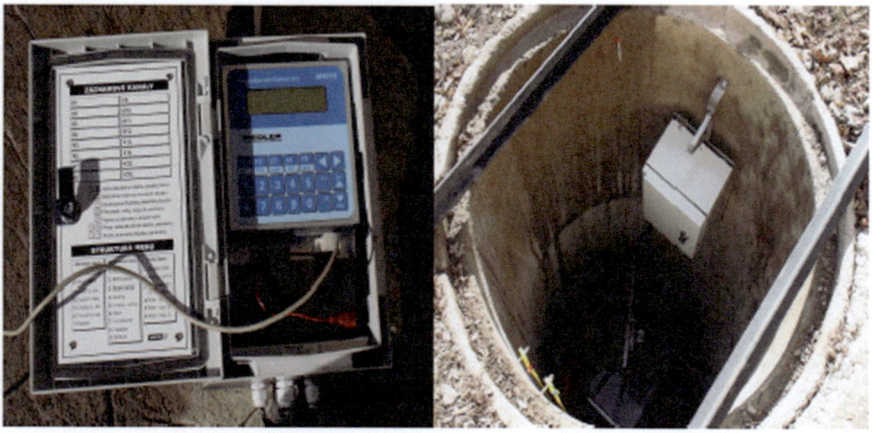

Fig. 3.20 Telemetric station M4016

The Solinst Levelogger allows you to easily export your data into spreadsheet format for Excel for further processing. The interval can be selected from 1 min to 99 h.

Telemetric Station M4016

The M4016 (Fig. 3.20), with full equipment, includes a universal data logger, a telemetry station with a built-in GSM module, a programmable control unit and a multiflowmeter in conjunction with ultrasonic or pressure-level sensors.

The telemetry station has 16 dynamically occupied recording channels for measuring and archiving pressure, levels, flow and other quantities. Recording channels have no fixed inputs. These can be analog, frequency or digital, and their number varies depending on the type of connection board. The assignment of the selected input to the measuring channel is done when setting the channel parameters. Measured signal values on individual inputs are stored in the data memory, which has a capacity of 2 MB and accepts up to 400,000 values, including date and time. The data memory also records extraordinary events—the receiving or sending of an SMS, occurrence of an error, network failure, etc. In normal mode with a constant power supply on the sensors, the averaged values measured over the entire archiving interval are stored at regular intervals. In standby mode, in which the central unit disconnects the sensor supply terminals from the battery and submerges itself, it stores only the results of a single measurement obtained at the end of the archive interval.

The M4016 is fitted with a built-in GSM module and can transfer measured values via GSM/GPRS communications to a server. In addition, units can send alerts or informative SMS messages. Low current consumption allows long-term powering of the entire device from the built-in rechargeable maintenance-free battery. This can be continuously recharged directly from an external power supply or from the solar panel. If the station is installed in field conditions without recharging the battery, it is possible to set the standby mode. In it, the station measures and tests input signals

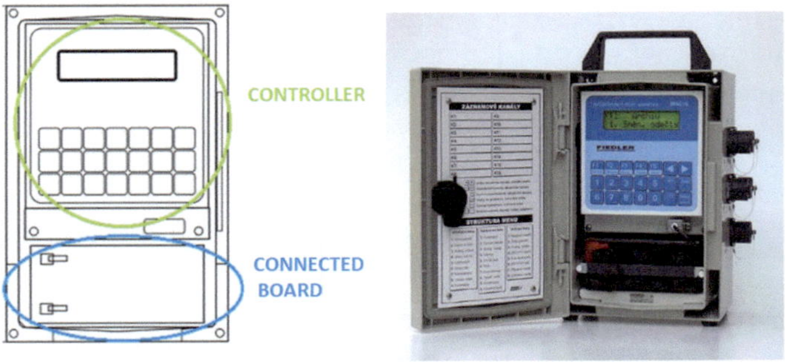

Fig. 3.21 Division of the telemetry station (Fiedler 2008)

only briefly at user-set intervals. Control electronics from input–output circuits, data memory and optionally with GSM module are enclosed in a 150 × 150 × 50 mm compact metal casting, which is equipped with a high-quality foil keyboard and backlit display on the front. A dual GSM antenna linked to the standard antenna connector is connected via a flat cable, with a series of terminals divided into groups according to the input signal type (Fig. 3.21).

The entire electronic part of the unit is enclosed in an IP66 metal casing to protect against external adverse and electromagnetic influences.

The station automatically logs into the GSM network after inserting a SIM card with activated data transfer. The SIM card must be unblocked before it is inserted into the station. It is used to transfer data to the Internet via GPRS. You must activate this service with your operator (e.g., by enabling MMS). When handling the SIM card, it is necessary to have the central control unit of the station disconnected from the supply voltage, which ensures the mechanical performance of the station itself. The SIM card hinged connector is accessible after removing the control unit from the station cabinet and is located under a stainless steel cap (Fig. 3.22) that protects the unit's interior against moisture penetration.

Terminals for connecting the input and output signals are located on the connection boards. With regard to the number and type of measurement being considered, the optimum station variant can be selected from the range of connection boards. The most commonly used type of connection plate is illustrated in Fig. 3.23.

Transferring measured and archived values from the M4016 station to the user's PC for further processing can be done in several ways (Fig. 3.24).

From the point of view of user comfort and operating costs, it seems more advantageous to use GPRS transmissions in a GSM network with data storage on the server when connected to the Internet.

When in March 2012 the B shaft was equipped with measurement devices, it was necessary to connect them to the M4016 control unit in shaft A (Fig. 3.25).

It should be noted that the telemetry station is the central unit for generating data from the devices used for this research. So all the devices are connected to this unit and sent to the Fiedler-Magr server.

Fig. 3.22 Preview into the control unit (Fiedler 2008)

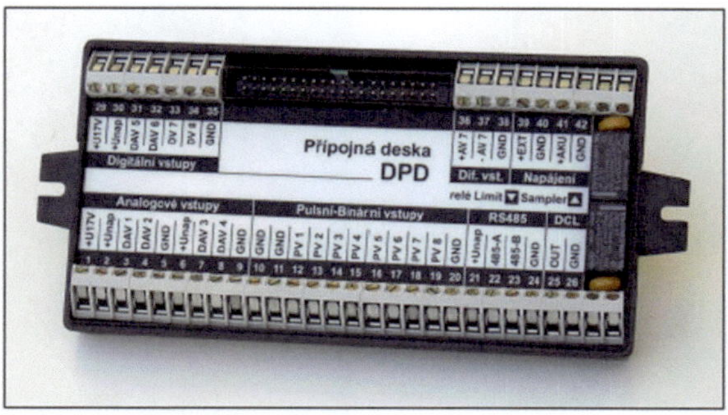

Fig. 3.23 Connection plate (Fiedler 2008)

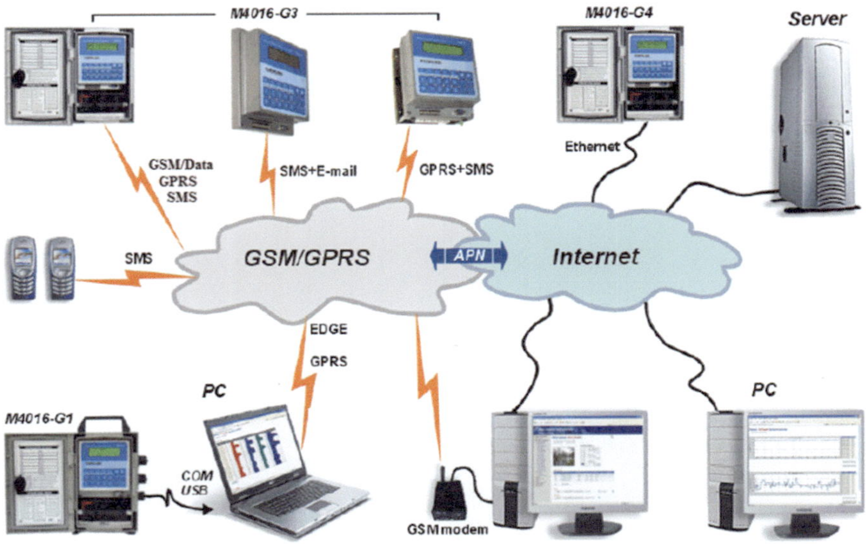

Fig. 3.24 Connections to M4016 Internet stations (Fiedler 2008)

Fig. 3.25 Connecting shaft B devices to the M4016 control unit in shaft A

Fiedler-Magr Server

Fiedler-Magr operates its own data server and leases data space for individual M4016 stations. For access to stored data, authorized users use a standard web browser, and no other visualization program is required. Part of the rental is also the creation of table and graphical visualizations of the measured data, supplemented by the daily statistical values of the individual measuring channels arranged in the monthly

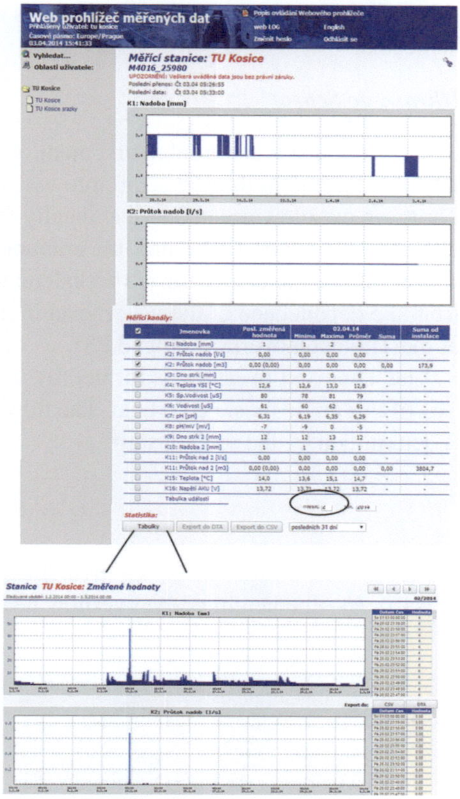

Fig. 3.26 Web browser of measured data from measuring devices on the Fiedler-Magr server

overview tables. The data for the selected time period can also be exported directly into Excel from the server database (Fig. 3.26).

From Fig. 3.26, it can be seen that the M4016 telemetry station that transmits the data to the Fiedler-Magr server is connected with the measuring instruments, by means of which the required quantities are measured. Specifically:

- ultrasonic sensor—the water level in the channel (vessel) is measured in [mm] and the flow rate in [$dm^3\ s^{-1}$], [m^3];
- YSI multiparameter probe—temperature [°C], conductivity [μS], pH;
- pressure sensor—measures the water level in the well in [mm];
- rain gauge—indicates the rainfall value in [mm].

The battery voltage is also shown in the Web browser of the measured data. As mentioned above, if the battery is weak, this warning information comes via the SMS operator.

More detailed information about these devices and measured parameters is provided below.

As the Web browser sends measurement data from the two wells, the measuring channels are marked from K1 to K15. It is possible to mark the squares on the left side, where the description of the channels we want to display the measured data for is located.

Ultrasonic Sensor

In this section, it is essential to first mention the metal measuring channel (Fig. 3.27), which is located in the wells and into which the rainwater flows from the roof. The Thomson overflow was made specifically for a particular well. It is used to place an ultrasonic sensor that measures the amount of water.

The ultrasonic sensor serves to calculate the flow rate. This calculation results from the dependence of the water level in the channel and operates on the principle of the Thomson overflow (Fig. 3.28).

The Thomson overflow is formed by two rectangular edges at a 90° angle. The axis of this angle must be vertical. The flow calculation using the Thomson overflow is determined by the relationship:

$$Q = K \cdot H^{\frac{5}{2}} \tag{3.1}$$

where Q—flow [dm^3 s^{-1}], K—hydraulic constant (for metal construction $K = 0.0146$), H—the height of the overflow beam [cm].

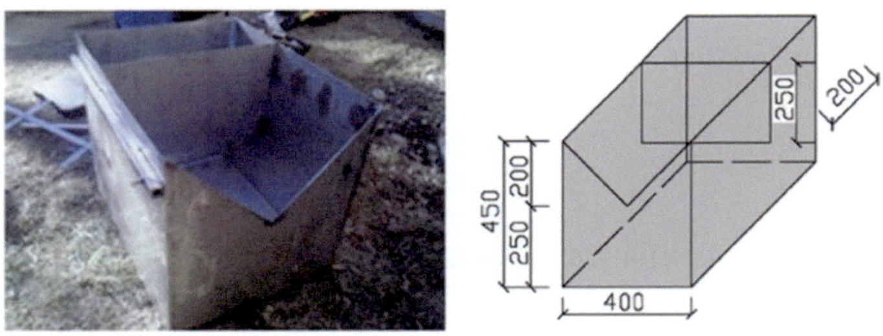

Fig. 3.27 Measuring channel—Thomson overflow

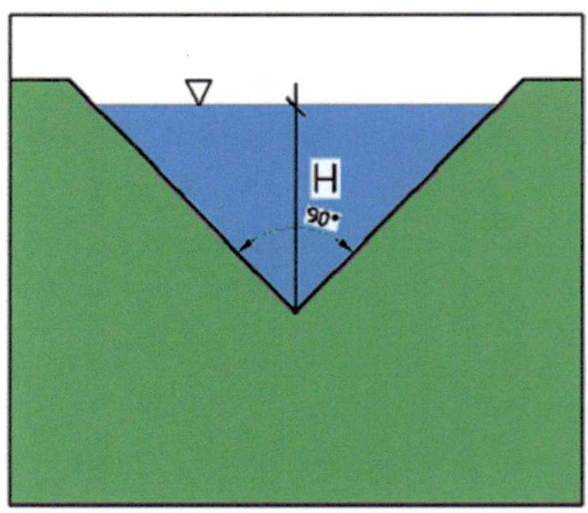

Fig. 3.28 Thomson overflow

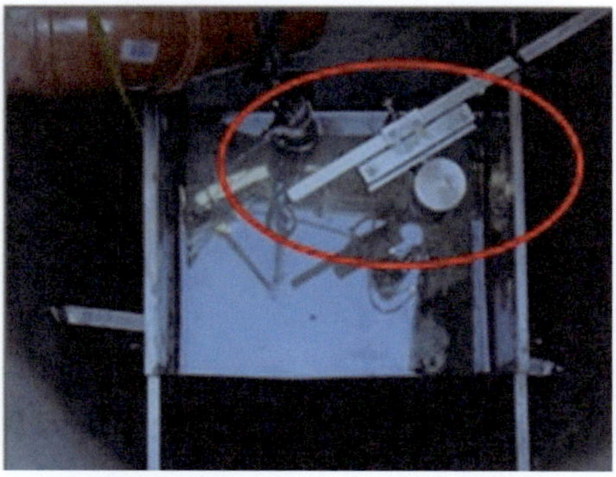

Fig. 3.29 Ultrasonic sensor—flowmeter

The height of the overflow beam for the specified heights ranges from 0 to 200 mm. The ultrasonic sensor also records the height of the rainwater in the channel.

The control unit records the data from the ultrasonic sensor, respectively the flowmeter (Fig. 3.29), which is located above the channel. It transmits the data at one-minute intervals, which are then transmitted to the Fiedler-Magr server (Fig. 3.30). The monitored data is then updated every 24 h, always at 5:00 in the morning.

Using the relationship (3.1) for the Thomson overflow calculation, flow data is recorded in $dm^3\ s^{-1}$ or m^3.

Together with the flowmeter, a multiparameter YSI probe is located above the measuring channel.

Multiparameter YSI Probe

The multiparameter YSI probe (Fig. 3.31) records the water quality parameters. The pH of rainwater, temperature of rainwater, conductivity as well as specific conductivity are monitored (Fig. 3.32).

The sensor portion of the probe is immersed in a channel in the rainwater drained into the shaft. As the sensor part can clog up, the probe must be calibrated at least once a month in pre-prepared dilutions for calibration. It is calibrated by connecting the probe to the EcoWatch software.

Pressure Sensor

At the bottom of the infiltration shafts, pressure sensors (float) of the LMP307 type (Fig. 3.33) are placed in a metal container, which continuously senses the height of the water level in the shaft. The pressure sensors are made of stainless steel with an IP68 degree of protection and are connected by a communication cable to the M4016 data telemetry unit, which transmits data to the server at one-minute intervals (Fig. 3.34).

The infiltration process, which can be tracked using measured data from a pressure sensor—data on the height of the water level in the well in mm—precedes the inflow of rainwater into the channel. Data on the water level in the infiltration shafts is used to calculate the infiltration capacity of the shaft.

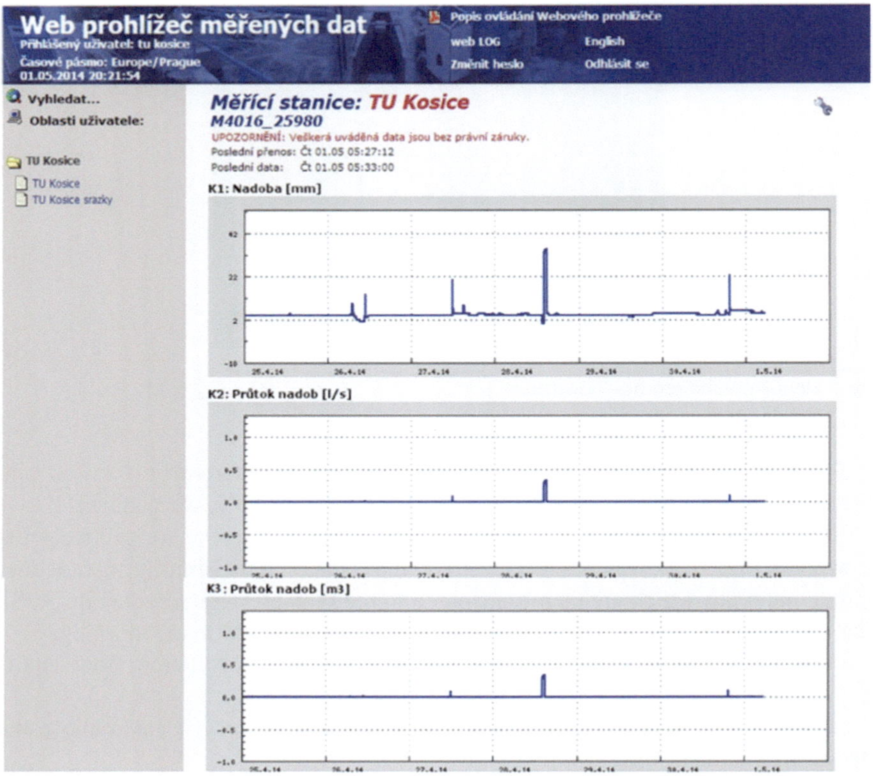

Fig. 3.30 Output from the Fiedler-Magr server for ultrasonic sensor data

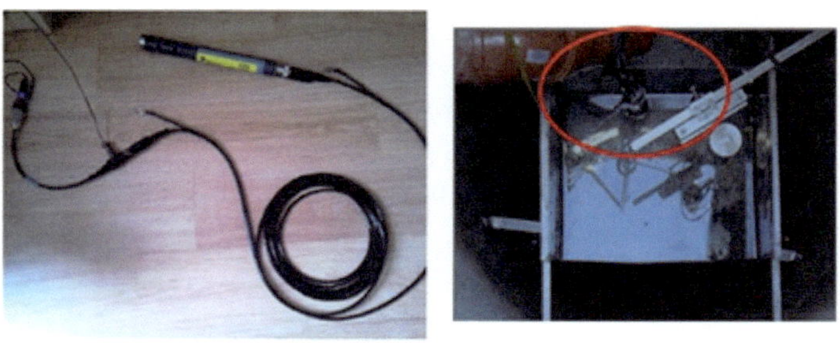

Fig. 3.31 Multiparameter YSI probe

The measured data from the above-mentioned devices is transmitted to the M4016 control unit and then via a communication module and stored to the Fiedler-Magr server in the form of a data host.

Figure 3.35 shows a diagram of the measuring instruments used and their location in the infiltration shaft.

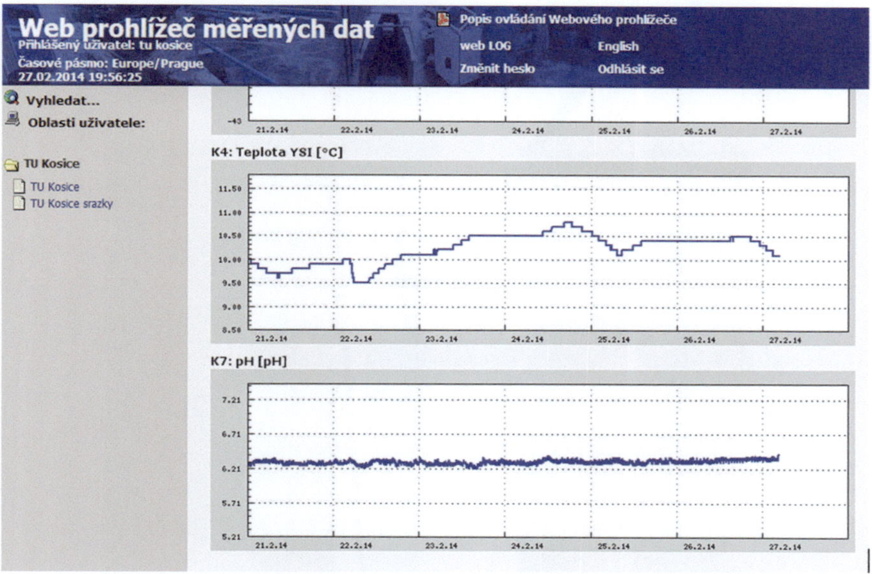

Fig. 3.32 Output from the Fiedler-Magr server for data from the YSI multiparameter probe

Fig. 3.33 Pressure sensor LMP307

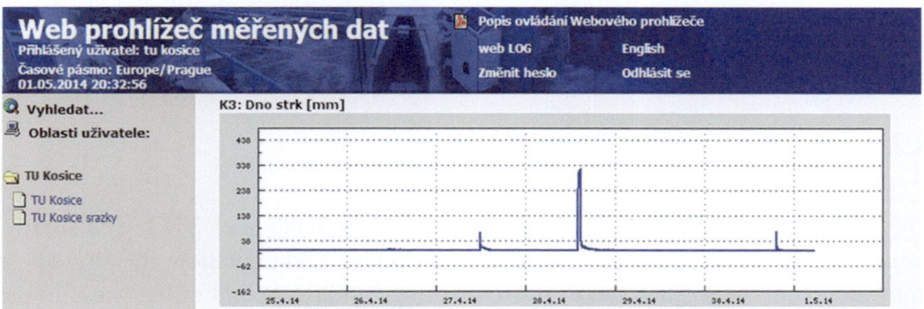

Fig. 3.34 Output from the Fiedler-Magr server for pressure sensor data

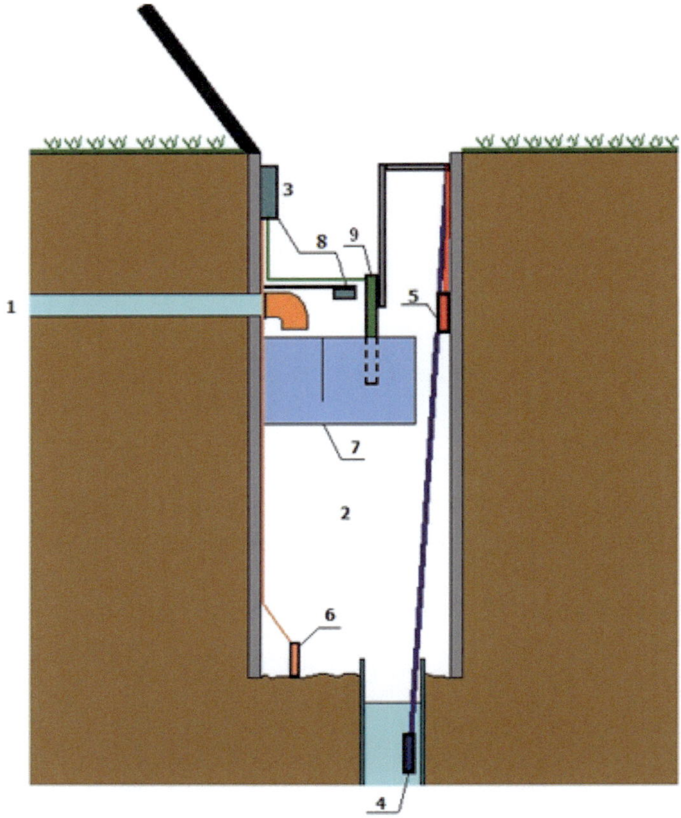

Fig. 3.35 Diagram of the shaft and the location of the instruments 1—pipe draining rainwater from the roof, 2—infiltration shaft, 3—data telemetry unit M4016, 4—levelogger, 5—barologger, 6—pressure sensor, 7—measuring channel, 8—ultrasonic sensor, 9—YSI multiparameter probe

All instruments are identical in both shafts; only the data telemetry unit M4016, the YSI multiparametric probe, the levelogger and the barologger are located only in shaft A.

Sections 2.3, 2.4, and 2.5 analyze these measurements.

3.2.2 Prešov—Model Location II

The first measurements in Prešov were launched in September 2012. The survey, similarly as in Košice, aims to monitor the amount of precipitation, the amount of rainwater from the surface runoff from the bridge and the height of the water in the filter shaft corresponding to the water level in the infiltration galleries.

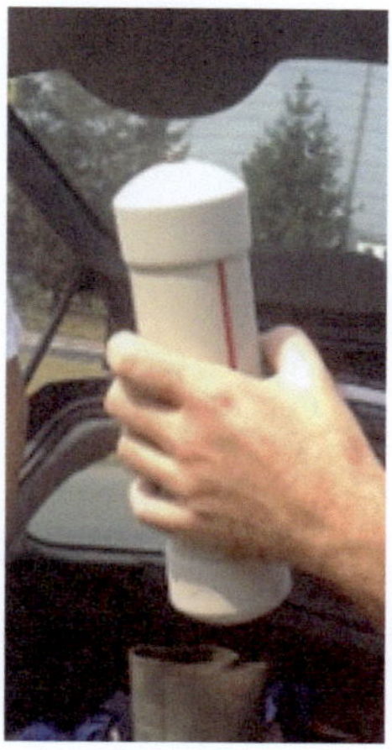

Fig. 3.36 Pressure sensor

Ultrasonic Sensor—Flowmeter

An ultrasonic sensor is placed in the filter shaft, a flowmeter that records the drained rainwater flow rate in $dm^3\ s^{-1}$ and m^3. The flowmeter is placed in a measuring channel.

Pressure Sensor

Using a pressure sensor (float) (Fig. 3.36), the height of the water in the infiltration galley can be determined. It is located in the inspection shaft (Fig. 3.37).

M4016 Telemetric Station and Fiedler-Magr Server

Measured data from measuring instruments is transmitted to the aforementioned M4016 telemetric station, which is located in the filter shaft, and then via the communication module, the data is stored to the Fiedler-Magr server in the form of a data host.

Figure 3.37 shows the measuring devices used and their location. The rainwater from the bridge is drained into the filter shaft. The filter shaft, the infiltration gallery and the inspection shaft are interconnected. The rainwater from the bridge is drained into the filter shaft (1), which serves to collect and deposit coarse and fine impurities. It is laid on a base plate with cast iron of class B125 and ventilation. In this filter shaft, there is also an M4016 control unit, a measuring channel with a Thomson overflow and a flowmeter. Subsequently, the rainwater enters the infiltration gallery (2), where the filtration process takes place. The infiltration gallery is covered with geotextile, against the penetration of backfill. The mixture contains a suitable compacted material. Similar to the infiltration shafts in Košice, the infiltration gallery is also ventilated. The inspection shaft (3) is equipped with a DN 600 concrete cover of class B125. In this research, it is also necessary to know the intensity and amount of precipitation; therefore, a rain gauge (4) is also located in the area.

The values obtained from the experimental measurements were the basis for our own research, presented in the following chapters.

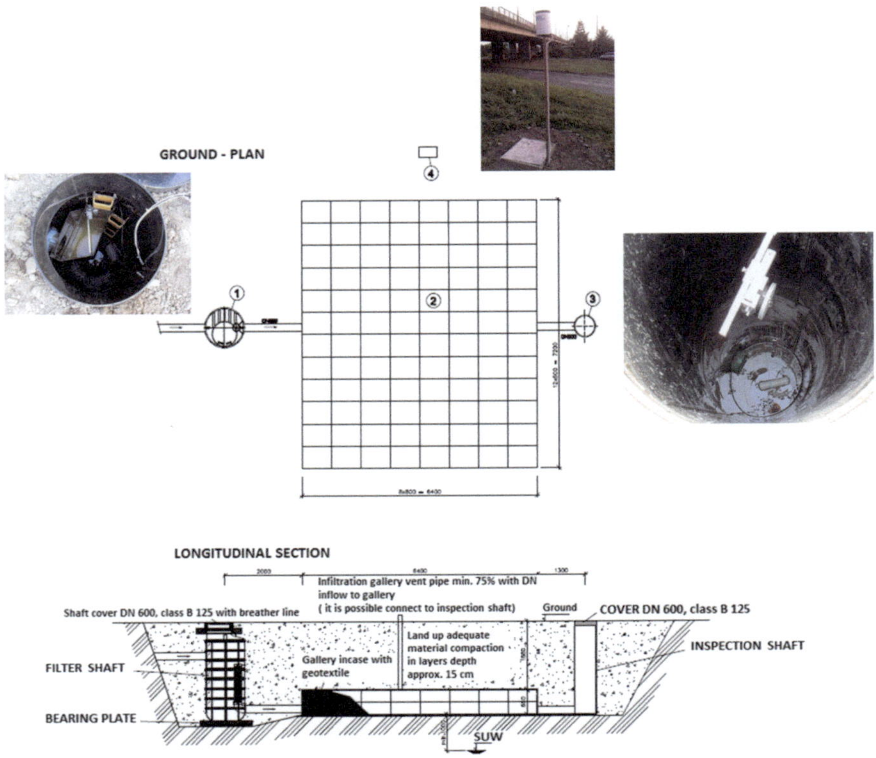

Fig. 3.37 Layout of terrain objects: 1—filter shaft, 2—infiltration gallery, 3—inspection shaft, 4—rain gauge, 5—ultrasonic sensor—flowmeter, 6—measuring channel, 7—pressure sensor, 8—data unit M40167

3.3 Terrain Measurements—Rainfall and Intensity of Rain

As mentioned in Sect. 2.2, the rain gauge is connected to a rainlogger, from which the data was manually retrieved using the Levelogger Solints software. The measurement method changed in August 2011 when the data was read automatically by the M4016 telemetry station. Measured data can be tracked online using the Fiedler-Magr Web browser as well as other parameters. The measured rainfall data is shown in Fig. 3.38.

Precipitation is subtracted at 10-min intervals, but for a more accurate overview of precipitation events, the interval changes to one minute from the start of the rain event. It is possible to import 30- and 60-min intervals of precipitation events.

A statistical table is used to show the daily minimum, maximum and average values (Fig. 3.39).

On each data page is also a button for exporting data into a file. The server can generate data for their further processing in MOST (DTA files) or in Microsoft Excel (CSV files).

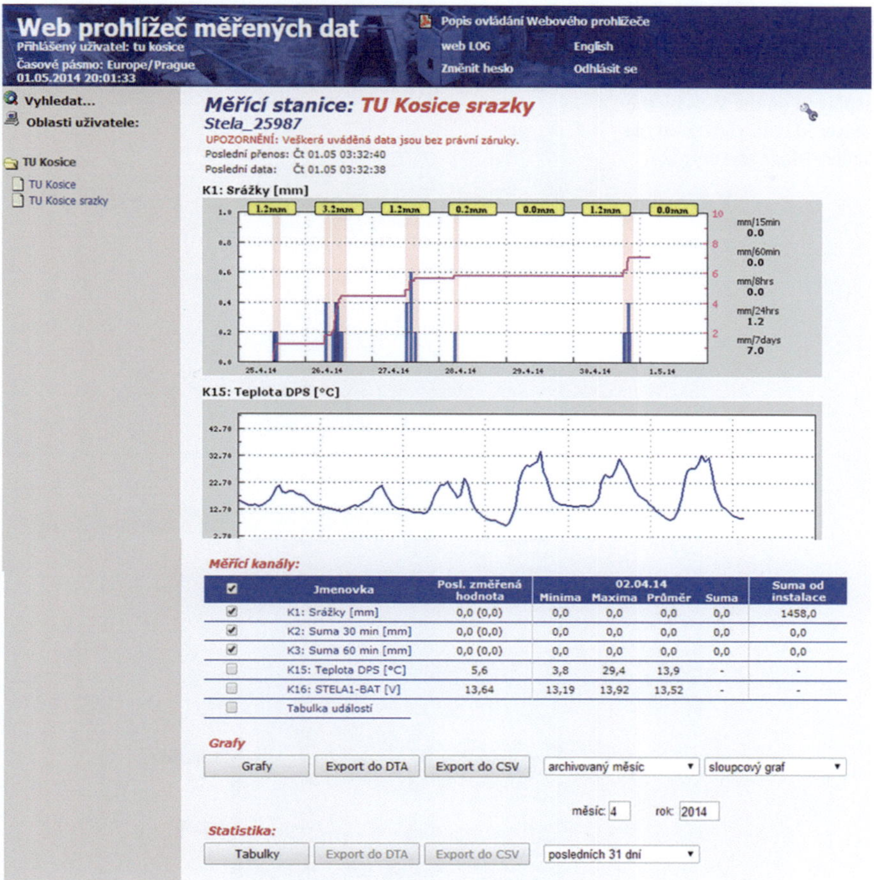

Fig. 3.38 Web browser of measured data on the Fiedler-Magr server

In the Šarišské Lúky area, the same rain gauge is located in the terrain as on the roof of the University Library on the campus of the Technical University in Košice. In this model area (Prešov), it is placed about 50 m from the infiltration gallery (Fig. 3.37). Data on precipitation totals can also be read from the server.

A table of maximum 30 and 60 min of precipitation was processed from the total rainfall from precipitation in Košice. The rainfall intensity calculation is given by the formula:

$$i = \frac{H_Z}{t_Z}[\text{mm}] \tag{3.2}$$

where H_Z—total precipitation [mm], t_Z—duration of rain [min].

Table 3.2 was available from the Slovak Hydrometeorological Institute (SHMI)

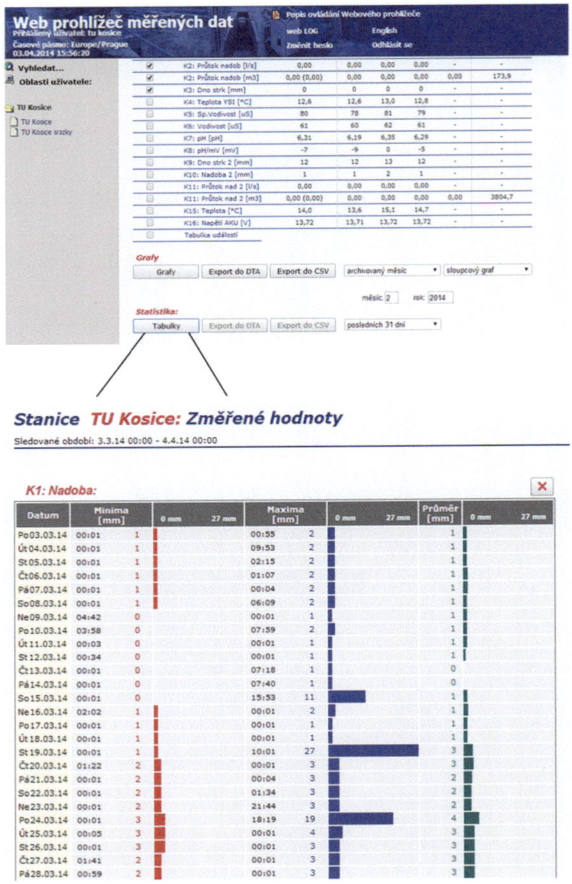

Fig. 3.39 Statistical table of minimum, maximum and average measured values displayed on the Web by the measured data viewer on the Fiedler-Magr server

in Košice. The duration of the precipitation compartments from the area of Košice-Barca.

We used this table to compare the graphically processed rainfall data and graphically processed maximum 30- and 60-min rainfall intensity. Relating the frequency: periodicity 5.0 means that rain events are repeated with a regularity of 5 times in 1 year, periodicity 2.0–2 times in 1 year, periodicity 1.0–1 times in 1 year, periodicity 0.5–1 times in 2 years, periodicity 0.2–1 times in 5 years, periodicity 0.1–1 times in 10 years, periodicity 0.05–1 times in 20 years and periodicity 0.033–1 times in 33 years. This suggests that the periodicity of 5.0 is the most undesirable. Precipitation totals [mm] are expressed in row *a*, and rain intensity [mm/min] is expressed in row *b*. The blue colors are the values we used for the comparison of precipitation totals (Sect. 3.1.1) and the red color for the comparison of rainfall intensities (Sect. 3.1.2).

Table 3.2 Duration of precipitation compartments

Frequency		Duration of precipitation compartments in minutes										
		5	10	15	20	30	40	50	60	90	120	180
5.0	*a* [mm]	4.9	6.2	7.0	7.5	8.2	8.5	8.7	8.9	9	9.1	9.3
	b [mm/min]	0.99	0.62	0.47	0.37	0.27	0.21	0.17	0.15	0.1	0.08	0.05
2.0	*a* [mm]	6.6	8.9	10.2	11	12	12.6	13	13.3	13.7	13.9	14.3
	b [mm/min]	1.32	0.89	0.68	0.55	0.4	0.32	0.26	0.22	0.15	0.12	0.08
1.0	*a* [mm]	8.2	11.0	12.4	13.6	14.9	15.8	16.4	16.8	17.7	18.1	18.7
	b [mm/min]	1.63	1.1	0.83	0.68	0.5	0.4	0.33	0.28	0.2	0.15	0.1
0.5	*a* [mm	9.4	12.8	14.8	16.2	17.9	19.1	19.9	20.4	21.5	21.8	22.1
	b [mm/min]	1.87	1.28	0.99	0.81	0.6	0.48	0.4	0.34	0.24	0.18	0.12
0.2	*a* [mm]	11.0	15.5	18	19.7	22	23.1	24.1	25.2	25.9	26.6	27.0
	b [mm/min]	2.2	1.55	1.2	0.98	0.73	0.58	0.48	0.42	0.29	0.22	0.15
0.1	*a* [mm]	12.1	17.4	20.4	22.5	25.2	26.6	27.6	28.7	29.9	31.0	31.3
	b [mm/min]	2.43	1.74	1.36	1.12	0.84	0.67	0.55	0.48	0.33	0.26	0.17
0.05	*a* [mm]	13.0	18.8	22.2	24.5	27.5	29.4	30.6	31.7	33.3	34.1	34.6
	b [mm/min]	2.61	1.88	1.48	1.22	0.92	0.73	0.61	0.53	0.37	0.28	0.19
0.033	*a* [mm]	13.9	20.2	23.9	26.3	29.5	31.7	32.4	33.6	35.4	36.3	36.9
	b [mm/min]	2.78	2.02	1.6	1.31	0.98	0.79	0.65	0.56	0.39	0.3	0.21

3.4 Determination of the Degree of Soil Permeability—The Filtration Coefficient

Infiltration of precipitation water into the rock environment depends largely on local conditions, especially the degree of its permeability. The basic hydraulic characteristic of such a measure is the filtration coefficient. Correct determination of its value is a key issue when assessing the infiltration of water into rocks, respectively soils. There are many different methods to determine it. In civil engineering, it is more common to determine the filtration coefficient using empirical calculations from grain curves. In hydrogeological practice, however, this basic hydraulic parameter is most precisely determined using terrain tests and hydrodynamic tests. Depending on the presence of groundwater, they may be either pumping or rising, and if water is not present in the examined rock environment, then pouring (infiltration) tests are conducted. The permeability of the rocks is discussed in more detail in Sect. 1.4.5.2.

3.5 Laboratory Test

This chapter is focused on the practical determination of grain categories of soil samples taken from two types of sites (both model localities), where climatic and hydrological parameters were observed to assess the possibility of precipitation infiltration in the rock environment to free (undefined) groundwater level. The first locality, as was mentioned, is on the Campus of the Technical University in Košice (Fig. 3.2), the second one at Šarišské Lúky (Fig. 3.7). Grain categories are determined either by a direct method or by indirect methods. The direct method—sieving on sieves with a certain mesh diameter—is the easiest way to determine grain size. This method cannot be used to determine the percentage of fine soil particles. Indirect methods are used in such a case.

The essence of the direct method test is to distribute the grains using a set of sieves to multiple grading classes with decreasing dimensions. The size of the holes and the number of meshes are chosen according to the type of draft and the required accuracy ( ).

The test procedure consists of washing and sieving under dry conditions. Since washing can change the physical properties of light aggregate, in such a case only dry sieving is used. The weight of the grains harvested on the different sieves relates to the original weight of the material. The total percentage of fallen grains by each sieve shall be given in numerical form and, if necessary, in graphical form (STN EN 933-1).

3.5.1 Testing Equipment

All test devices must comply with the general requirements stated in standards, specifically prEN 932-5: 1999 (STN EN 933-1).

Testing equipment includes (STN EN 933-1)

- test sieves, with square holes complying with EN 933-2 and meeting the requirements of ISO 3310-1 and ISO 3310-2;
- bottom and cover adapted to the sieves;
- an air-operated dryer controlled by a thermostat to keep the temperature 110 ± 5 °C or other non-grinding aggregate suitable for grain drying;
- washing equipment;
- weighing instruments with an accuracy of 0.1% of the test sample weight;
- bowls, brushes;
- automatic sorting (at your discretion).

3.5.2 Test Procedure

The taken samples must be reduced in accordance with prEN 932-2 to obtain the required number of test samples. For a sample with a high fine grain size, a sample humidification to reduce segregation and loss of dust (STN EN 933-1) may be required prior to the reduction process.

The weight of each test sample shall correspond to the requirements in Table 3.3, for aggregates weighing between 2000 and 3000 kg/m^3 (STN EN 933-1).

The reduction of samples must allow for obtaining samples of a mass greater than the minimum, but without a predetermined value. The test sample is dried at a temperature (110 ± 5)°C until a steady state is obtained, allowed to cool, and the result weighed and recorded as M1. The fixed weight is the result of two successive weighings after separate drying for at least 1 h, which does not differ by more than 0.1%. For some types of aggregate, drying at 110 °C causes a high grain bonding that prevents grain separation during the washing and/or sowing process. For such aggregates, the procedure specified in STN EN 933-1 (normative) should be used.

In the framework of the research, there were three soil samples, namely two samples from model area I (Košice) and one sample from model area II (Prešov), which are mentioned in the chapter on field tests.

Samples from study areas (Fig. 3.40) were delivered to the GEO Slovakia Laboratory, Ltd, in Košice (Fig. 3.41) to determine the filtration coefficient.

The supplied samples were pounded and marked (Fig. 3.42). The weight of these samples was 5 kg. This weight was determined by the methodology.

The test sample is used as a whole for one test. The process of determining the grain curve beginning from the sample's arrival at the laboratory until determination of the filtration coefficient is shown in Fig. 3.43.

Table 3.3 Mass of test pieces of ordinary aggregate (STN EN 933-1)

Grain size of aggregate *D* (maximum) [mm]	Test sample weight (minimum) [kg]
63	40
32	10
16	2.6
8	0.6
≤ 4	0.2

Note 1 For aggregates with a different grain size, the minimum mass of the sample may be interpolated from the masses shown in Table 3.1
Note 2 If the mass of the sample does not correspond to the data in Table 3.1, the grain size obtained does not correspond to this standard and this must be stated in the test report
Note 3 For aggregates with a bulk density of less than 2000 kg/m^3 or more than 3000 kg/m^3 (according to prEN 1097-6), the corresponding correction for the mass of the test sample given in Table 3.1, on the basis of a weight-to-weight ratio, in order to obtain a test sample corresponding to approximately the same volume as for conventional aggregate

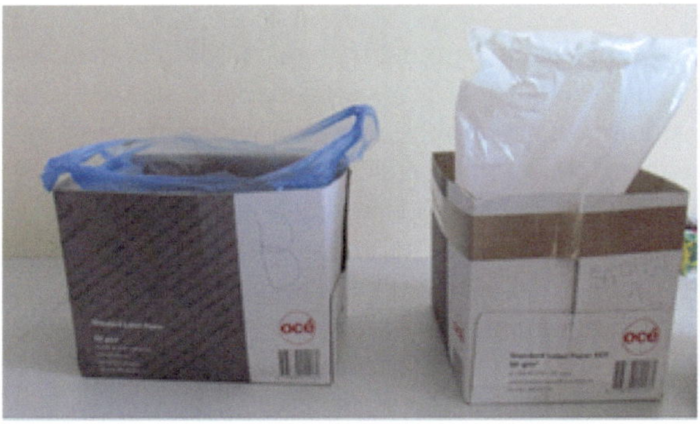

Fig. 3.40 Samples from shafts A and B—Košice

Fig. 3.41 Laboratory GEO Slovakia Ltd Košice

Fig. 3.42 Samples before drying and washing

Fig. 3.43 Workflow diagram for determining the filtration coefficient in the laboratory

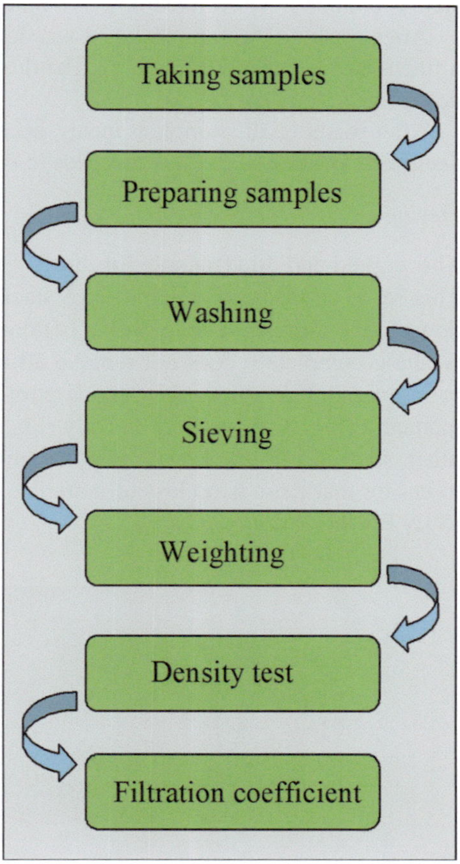

The procedure for grain testing and subsequent determination of the filtration coefficient comprises five steps. It is divided into two parts: a sieve and a dense part. The sieve includes washing, sieving and weighing. This procedure is described in detail in the following section.

Washing

The test sample is placed in the vessel, and the necessary amount of water is added to cover the aggregate. The soak time of 24 h will ease the lumps. A dispersing agent may also be used. The test sample is shaken sufficiently intensively to achieve complete separation and formation of a fine grain suspension (STN EN 933-1).

Both sides of a 0.063 mm sieve to be used for this test only are moistened, and a screening sieve (e.g., with 1 mm or 2 mm holes) placed on the sieve is used. The sieves are placed such that the suspension passing through the sieve can be removed or, if necessary, taken into a suitable container. The container content is placed on the top screen. The washing is continued until the water passing through the 0.063 mm sieve is clear (STN EN 933-1).

The 0.063 mm site residue is dried at a temperature (110 ± 5)°C until a steady state is obtained. Allow to cool, weigh and record the result as M_2.

After washing the samples, dust was left (Fig. 3.44), and the rest was soil with a fraction of 8–0.1 cm (Fig. 3.45). The dust particles are not complete; the rest are determined by a density test.

After washing, the sample is mealy, because the fine fraction of the soil has been excluded. The rest of the soil proceeds to the next process, which is sieving.

Sieving

The washed and dried sample (or directly dry sample) is placed on a set of sieves. This set is made up of a number of stacked meshes and placed from the top to down to the diminutive hole, with a bottom and a cover. Experience has shown that washing completely does not remove all the fine grains. It is therefore necessary to insert a 0.063 mm mesh in the control socket set. The sieving set is shredded manually (Fig. 3.46) or mechanically (Fig. 3.47). Then one sieve is taken from the other, starting with the largest aperture sieve, and each sieve is shredded manually to ensure that there is no loss of material, using, for example, the bottom or cover (STN EN 933-1).

Fig. 3.44 Dust from samples

Fig. 3.45 Soil with a fraction of 8.0–0.1 cm

Fig. 3.46 Hand sieving

The mechanical oscillator used in the laboratory was equipped with sieve openings: Ø 4; 2; 1; 0.5; 0.25; and 0.1 mm. Any material that falls through each site will be stuck to the next sieve before proceeding to sieving on this sieve. Sieving can be considered to be complete if the balance does not change more than 1.0% during a 1-min sieving. The sieving time is not given in the standard. It depends on the type of soil. The sieving time of the test sample in the laboratory was determined for 30 min. A suitable sieving time is determined by checking by means of manual scrolling.

Fig. 3.47 Mechanical oscillator

To avoid overfilling, the residual on each site is obtained at the end of the sifting, expressed in grams, and it should not exceed (STN EN 933-1):

$$\frac{A \cdot d^{\frac{1}{2}}}{200} \tag{3.3}$$

where A—area of sieve [mm^2], d—dimensions of sieve openings in [mm].

If one of the balances exceeds this quantity, one of the following procedures (STN EN 933-1 2009) must be used:

- the residual is divided into smaller parts than is the prescribed maximum and these parts are sieved one on the other;
- the portion of the sample overlapping with the largest holes is divided by means of a sampler or square, followed by sifting of the reduced sample, taking into account the reduction of the sample.

If the sieving process is completed, the weighing of the soil from the mechanical oscillator is continued.

Weighing

First, the material from the sieve with the largest dimensions is weighed (Fig. 3.48), and the mass R_1 is recorded.

The same operation is performed on the sieve with smaller holes, and the mass R_2 is recorded. This continues with the same operation on each sieve that is in the set to obtain the weight of the various trapped portions of the material, and the masses of R_3, R_4, … R_1, … R_n (STN EN 933-1) are recorded.

The units of weight obtained are converted to percentages, on the basis of which a grain curve is plotted, and the value of the filtration coefficient is determined.

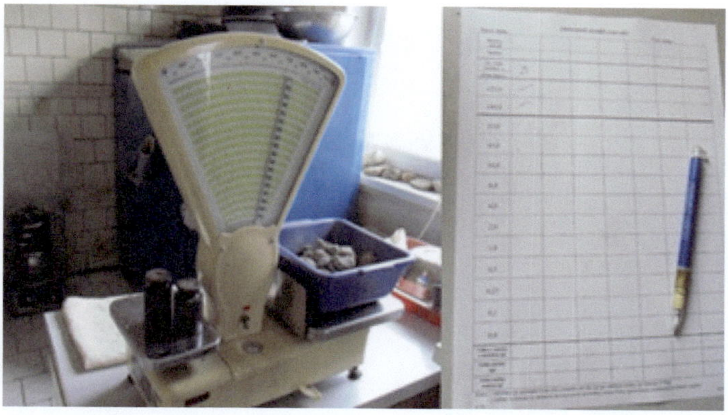

Fig. 3.48 Weighing of samples

Density Test

The density test is performed to complement and close the graph of the grain curve of the soil, from which the filtration coefficient is determined.

The density test is made of clay soils, where the grains are less than 4 mm or from the overflow of the washed samples (dust drop, grains <0.1 mm). The sample must be mixed during the initial measurement. The density is monitored by a densimeter at time intervals. The measurement time is carried out after 1.5 min, 15 min, 60 min, 4 h and 24 h, and the obtained value is recorded. This measurement determines how the sample sediments (Fig. 3.49).

Measurements also monitor a temperature that should not have high deviations, so the laboratory must be equipped with air conditioning.

The densimeter (Fig. 3.50) or the aerometer used in the densitometry test is designed for the analysis of the soil. It is a submerged body, mostly in the form of a bank, with a protruding stem in which a scale is placed. The depth of the dive in the liquid is dependent on the density of the liquid. Densimeters are used to measure the density of liquids on the basis of Archimedes' principle. Metal balls are closed at the bottom to accommodate the weight of the bank relative to its volume. On the tapered part of the tube is the scale in density units (kg m^{-3}). When measuring the density, the densimeter is sunk into the liquid so that only the stem of the bank with the scale rises out of the liquid.

The whole screening process, as well as the density test, serves for calculating the filtration coefficient, in addition to determining the fraction of the soil. When calculating the filtration factor, the size of the gravel grain of 10 and 60% is important (Sect. 1.4.5.1). The filtration coefficient is given in m s^{-1}.

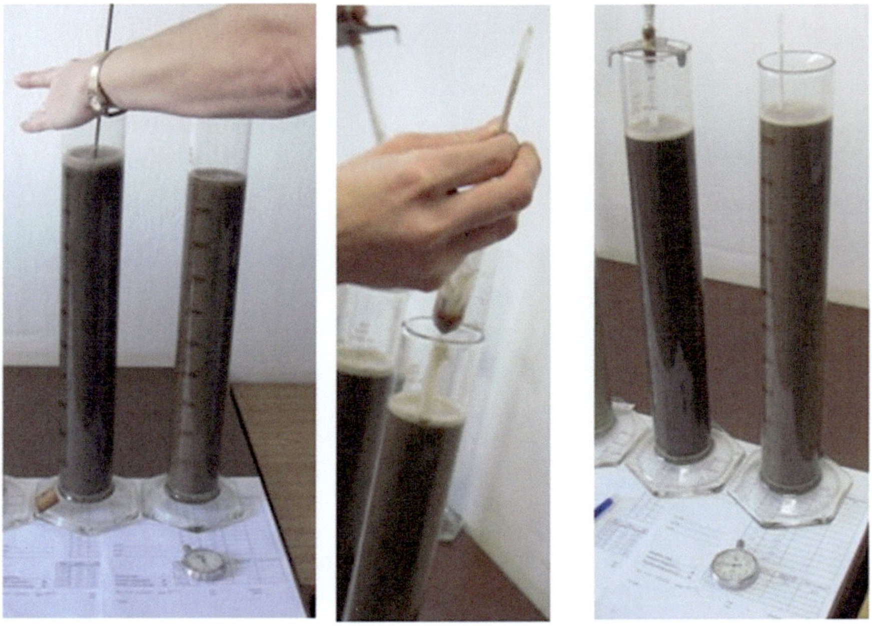

Fig. 3.49 Density test

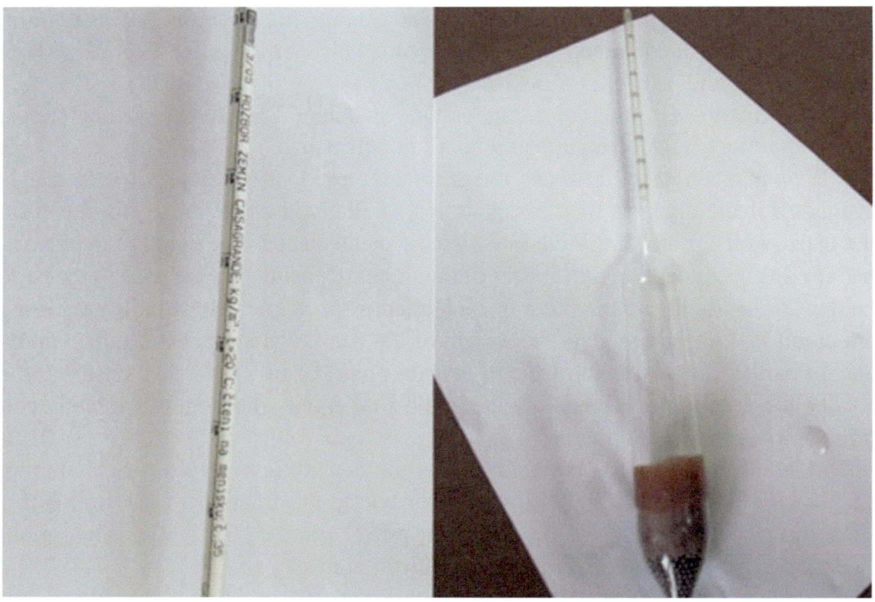

Fig. 3.50 Densimeter

3.5.3 Empirical Relations for Calculating the Filtration Coefficient

The filtration coefficient can be calculated in several ways. The first possibility to calculate the filtration coefficient is given in ČSN 75 9010.

$$Q_I = \frac{V}{T_E}\ [\mathrm{m^3\ s^{-1}}] \tag{3.4}$$

$$A_I = \pi \cdot r^2\ [\mathrm{m^2}] \tag{3.5}$$

or

$$A_I = a \cdot b\ [\mathrm{m^2}] \tag{3.6}$$

$$k_f = \frac{f \cdot Q_I}{A_I}\ [\mathrm{m\ s^{-1}}] \tag{3.7}$$

where Q_I—infiltration flow [$\mathrm{m^3\ s^{-1}}$], T_E—emptying time [s], V—retention volume [$\mathrm{m^3}$], f—surcharge factor (for calculation: $f = 2.0$), k_f—filtration coefficient [$\mathrm{m\ s^{-1}}$], A_I—infiltration area [$\mathrm{m^2}$], r—radius of the infiltration shaft in model area I (for calculation: 0.5 m), a, b—the dimensions of infiltration gallery in model area II (for calculation: $a = 8 \times 0.8 = 6.4$ m; $b = 12 \times 0.6 = 7.2$ m).

The second option is to calculate the coefficient of filtration by empirical equations given by Geoton Ltd published in the final report of the Geological Survey of Sabinov (2012). Calculation of the infiltration coefficient was derived from the following formulas:

$$T_E = \frac{R_P}{R_T} \cdot D_R\ [\mathrm{s}] \tag{3.8}$$

$$A_I = \frac{k_{0.9}}{k_f \cdot T_E}\ [\mathrm{m^2}] \tag{3.9}$$

The final formula for calculating the filtration coefficient then looks like this:

$$k_f = \frac{R_T \cdot k_{0.9}}{R_P \cdot A_I \cdot D_R}\ [\mathrm{m\ s^{-1}}] \tag{3.10}$$

where k_f—filtration coefficient [$\mathrm{m\ s^{-1}}$], T_E—emptying time [s], R_P—15-min rain at a periodicity of 0.2 (for calculation: 18.10^{-3}) [m], R_T—highest monthly precipitation total per year (for calculation: 0.097; 0.1288) [m], D_R—number of days in month with higher average rainfall converted to seconds [s], A_I—infiltration area [$\mathrm{m^2}$], c_{09}—drainage coefficient (reduced precipitation over 15 min of rainfall on a paved surface) [$\mathrm{m^3}$].

$$k_{09} = A_R \cdot R_{15} \cdot 0.9\ [\mathrm{m}^2] \tag{3.11}$$

where A_R—roof area (for calculation: 33.7 × 15.10) [m^2], R_{15}—precipitation in 15-min time intervals for the measured day [m], 0.9—drainage coefficient (for calculation: 0.9).

The highest monthly rainfall (R_T) for 2012 was 128 mm in June and 97 mm in May. As research began in August 2011, the 2012 value was considered for the year and similarly assumed was 2014, where 97 mm was considered.

The number of days in the month with the highest average rainfall (D_R) was determined from the highest monthly averages for the year, divided by the number of days in that month. For both May and July, this is 31 days. This calculation determined the value that was considered for determining the number of days with the highest precipitation average total. Of course, this and higher value were taken into account in determining the number of days.

$$D_R\,\text{July 2012} = \frac{128.8}{31} = 4.15\ [\mathrm{mm}] \tag{3.12}$$

$$D_R\,\text{May 2013} = \frac{97}{31} = 3.13\ [\mathrm{mm}] \tag{3.13}$$

In July 2012, the resulting value is 4.15 mm. This value was exceeded 8 days in the given month, representing 691,200 s. The value in May 2013 is 3.13 mm. The number of days that are above this value is 7 and thus 604,800 s.

The results of these values are reviewed in the chapter.

3.6 Dimensional Analysis—A Mathematical–Physical Model of Emptying Time Determination

The mathematical–physical model was created in the framework of this research based on the dimensional analysis, the methodology of which is described in Sect. 1.4.6.

The relevant variables influencing the emptying time of infiltration were selected as follows:

- rain intensity i [m s^{-1}];
- infiltration area A_I [m^2];
- emptying time T_E [s];
- specific weight/density of soil ρ [kg m^{-3}];
- filtration coefficient k_f [m s^{-1}].

The complete physical equation expressing the dependence of selected relevant dimensional variables with different dimensions has the following form according to (2.44):

$$\varphi(i, A_I, T_E, \rho, k_f) = 0$$

On the basis of the dimensional diversity of the relevant variables, they proceed according to Eq. (2.45):

$$\pi_i = i^{x_1} \cdot A_I^{x_2} \cdot T_E^{x_3} \cdot \rho^{x_4} \cdot k_f^{x_5}$$

The dimensional matrix A for the basic units of selected physical quantities has the number of rows $m = 3$ and the number of columns $n = 5$. At the rank of the matrix $h = 3$ and the number of relevant variables $n = 5$, it is possible to create a total of $i = n - m$, two-dimensional arguments. According to the relationship (2.48), the dimension matrix looks as follows:

	i	A_I	T_E	ρ	k_f
kg	0	0	0	1	0
m	1	2	0	−3	1
s	−1	0	1	0	−1

The number of unknown physical quantities $x_i > m$ (number of dimensions); therefore, unknown quantities cannot be uniquely determined. Thus, the rectangular dimensional matrix (2.48) was divided into two parts. The created square matrix has the number of columns n and the number of rows m, with the columns of the matrix being selected so that the matrix determinant is nonzero ($\Delta \neq 0$). According to the relation (2.51), the vector of unknown quantities x_i was also divided.

$$\begin{Vmatrix} 0 & 0 & 1 & 0 \\ 1 & 2 & -3 & 1 \\ -1 & 0 & 0 & -1 \end{Vmatrix} \cdot \begin{Vmatrix} x_1 \\ x_2 \\ x_3 \\ x_4 \end{Vmatrix} = (-1) \cdot \begin{Vmatrix} 0 \\ 0 \\ 1 \end{Vmatrix} \cdot \|x_5\|$$

The matrix by relation (2.51) has been transformed into a system of four linear equations with five unknowns.

$$x_3 = 0$$
$$x_1 + 2x_2 - 3x_3 + x_4 = 0$$
$$x_1 - x_4 = -x_5$$

The choice of the unknown variable was made according to the relationship (2.55) twice:

$$1.\,\text{option}: x_1 = 0, x_5 = 1$$
$$2.\,\text{option}: x_1 = 1, x_5 = 0$$

Table 3.4 Obtained independent vectors

	x_1	x_2	x_3	x_4	x_5
π_1	0	1/2	0	1	1
π_2	1	0	0	−1	0

By solving the system of linear equations for conditions according to relation (2.55), two independent vectors were obtained (Table 3.4):

This solution [after adding information about the values $x_1, \ldots, x_5$ to the relation (2.45)] answers a complex dimensionless argument in the form:

$$\pi_1 = A_f^{1/2} \cdot k_f \cdot T_E$$

The second dimensionless argument is in the form:

$$\pi_2 = i \cdot k_f^{-1}$$

The dimensionless shape of the function of the time-counting function can be written in the form according to the relation (2.46):

$$\varphi(\pi_1, \pi_2) = 0$$

respectively

$$\varphi\left(\frac{T_E \cdot k_f}{\sqrt{A_I}}, \frac{i}{k_f}\right) = 0 \tag{3.14}$$

The dimensionless argument π_1 contains the variable T_E; therefore, this argument can be expressed as a function of the π_2 argument in the form:

$$\pi_1 = \varphi(\pi_2) \tag{3.15}$$

From the measured values of the relevant variables in the model area (Table 4.12), the actual function of $\pi_1 = \varphi(\pi_2)$ and the scattering array of the points of these dimensionless arguments in the logarithmic coordinates (Figs. 4.11 and 4.12) can be depicted.

Assuming the linear dependence of these dimensionless arguments π_1, π_2 (valid for the exponential dependence displayed in logarithmic coordinates), the relation (3.15) between the independent variable π_2 and the dependent variable π_1 can be described by an exponential equation in the form:

$$\pi_1 = A.\pi_2 \tag{3.16}$$

which was transformed by the logarithm on the line equation

$$\log \pi_1 = \log A + B \cdot \log \pi_2 \tag{3.17}$$

The calculation of the locating constant A (giving the distance of the intersection of the regression line to the coordinate π_1 from zero) and the regression coefficient B (indicating the regression line directive) was determined by the least squares method,

$$a = \frac{\sum x_i^2 \cdot \sum y_i - \sum x_i \cdot \sum x_i \cdot y_i}{n \cdot \sum x_i^2 - \left(\sum x_i\right)^2} \tag{3.18}$$

$$b = \frac{n \cdot \sum x_i \cdot y_i - \sum x_i \cdot \sum y_i}{n \cdot \sum x_i^2 - \left(\sum x_i\right)^2} \tag{3.19}$$

where

$$\sum x_i = \sum \log \pi_2,$$
$$\sum y_i = \sum \log \pi_1,$$

than $a = \log A \quad \Rightarrow A = 10^{a}, b = B$.

Using the dimensional analysis, the mathematical–physical model of the emptying time has the following form of dependence of the relevant dimensionless variables after Eq. (3.16)

$$T_E \cdot \frac{k_f}{\sqrt{A_I}} = A \cdot \left(\frac{i}{k_f}\right)^B$$
$$T_{vs} \cdot \frac{k_f}{\sqrt{S_i}} = A \cdot \left(\frac{i}{k_f}\right)^B \tag{3.20}$$

After the adjustment, the following relationship was obtained:

$$T_E = A \cdot A_I^{-\frac{1}{2}} \cdot k_f^{(-1-B)} i^B \tag{3.21}$$

The relationship is universally valid for the emptying time calculation, but for each case it is necessary to calculate the new A and B coefficients.

References

Emergency National Center (2014). [online] [cit. 18.1.2018]. Online http://www.emergency-slovakia.sk/kos-zzs-2.html?page_id=144

Fiedler J (2008) M4016-G, uživatelská příručka verze 1.10. Registrační a řídící jednotka, Telemetrická stanice, Průtokoměr

Technical report (2012) I/18 Šarišské Lúky oprava mosta č.18-449, Prodosings. r. o., Bardejovská 13, 080 06 Ľubotice

Chapter 4
Investigation of Rainwater Infiltration in Model Areas

Abstract This chapter presents research results in model areas I and II. The results include graphical processing of total rainfall and rainfall intensity in Košice, as the tabled rainfall intensities from Slovak Hydrometeorological Institute (SHMI) (Table 3.2) were available only for this area. More detailed results are given in the annexes: *C* (relationship between rainfall and infiltration), *D* (rainfall total) and E (rainfall intensity). The chapter also presents the filtration coefficients obtained from the hydrodynamic tests, from the grain curves and the calculations based on empirical relations. In the following, a developed model to calculate the emptying time of the infiltration device is verified.

4.1 Evaluation of Hydrological Parameters—Precipitation

As was mentioned above, data (Table 3.2) on rainfall for the Košice-Barca rain gauge station was available from the SHMI branch office in Košice. This rainfall data, marked with a blue colour (marked also by letter a for the corresponding periodicity), were used to prepare the graphs comparing the measured rainfall totals and the table rainfall data for the corresponding periodicity. The measured values are from a rain gauge located at the University Library in Košice.

4.1.1 Precipitation Totals

In this chapter, the charts of precipitation totals and their comparison with the table values are processed. Table 4.1 and Fig. 4.1 serve as examples.

We used data from rainfall events to verify table values for the individual periodicities from SHMI (1968).

M. Zeleňáková et al., *Rainwater Infiltration in Urban Areas*,
Water Science and Technology Library 89,
https://doi.org/10.1007/978-3-030-34698-0_4

Table 4.1 Values of measured precipitation and table precipitation at periodicity 5.0 from the rainfall event of 4.6.2013 (16:10–19:10) in Košice

Time [min]	5	10	15	20	30	40	50	60	90	120	180
Measured precipitations [mm]	2	4.6	8.2	10	11.4	11.6	12	12.2	12.8	15.2	16
Table precipitations [mm]	4.9	6.2	7	7.5	8.2	8.5	8.7	8.9	9	9.1	9.3

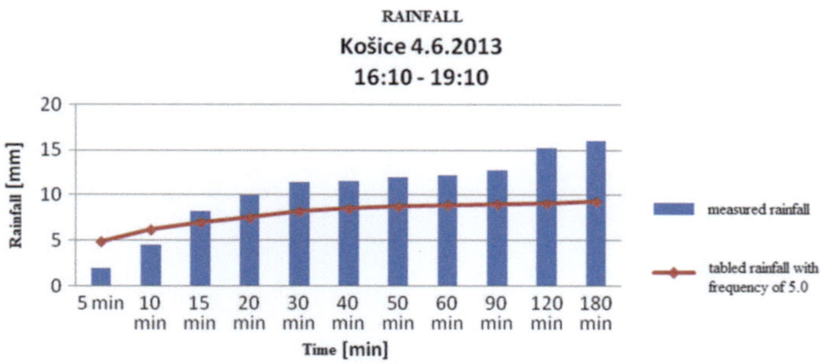

Fig. 4.1 Graph showing measured rainfall values and table values at periodicity 5.0 from the rainfall event of 4.6.2013 (16:10–19:10) in Košice

4.1.2 Rainfall Intensity

Rainfall events were recorded for each day and the total rainfall for that day. Measurements of precipitation totals for 1-min, 30- and 60-min intervals are also recorded. The calculation of the maximum intensity at the 30- or 60-min interval was according to Eq. (3.2).

From Tables 4.2 and 4.3, containing a maximum of 30- and 60-min intensities for 2011–2014, and Table 3.2, the duration of precipitation compartments given by the SHMI (1968), a graph was created (Fig. 4.2) which shows the relationship between the measured and table rainfall intensities.

Values that exceeded 30- and 60-min intensities at the periodicity of 5.0 from Table 4.7 are highlighted in red in Tables 4.2 and Table 4.3.

4.2 Evaluation of the Hydrogeological Parameter—Filtration Coefficient

4.2.1 Determination of the Filtration Coefficient by the Hydrodynamic Test

For precise determination of the filtration coefficient, the in situ hydrodynamic test methods described in Sect. 1.4.5.2 are the most appropriate. However, there was no possibility of their determination from the category I of quality of documents set by Jetel (1985), nor was it possible to use the archival results from older research work.

In this case, the most appropriate method emerges from the II. qualitative category, with the possibility of estimating approximate (comparative) parameters derived from the specific yield with an indirect estimate of strictly hydraulic parameters, which also predates the filtration coefficient.

Table 4.2 Maximum rainfall in 30 min—June 2012

June 2012	Rainfall [mm]	Number of measurements		Max. intensity (30 min)			Min. intensity (30 min)		
		1 min	30 min	Time [day]	Max. rainfall [mm]	Max. intensity [mm/min]	Time [day]	Min. rainfall [mm]	Min. intensity [mm/min]
1.6.2012	0.8	4	4	8:30	0.2	0.0066667	8:30	0.2	0.0066667
4.6.2012	11.4	30	2	23:30	7.2	0.24	0:00	4.2	0.14
5.6.2012	10	50	14	1:00	2.8	0.0933333	1:30	0.2	0.0066667
6.6.2012	0.2	1	1	1:30	0.2	0.0066667	1:30	0.2	0.0066667
7.6.2012	4.2	10	2	13:00	3.8	0.1266667	13:30	0.4	0.0133333
9.6.2012	7.2	29	4	2:00	4.4	0.1466667	21:30	0.2	0.0066667
10.6.2012	5.6	28	4	2:30	2.8	0.0933333	1:30	0.2	0.0066667
11.6.2012	6.6	20	4	11:30	4.8	0.16	1:30	0.2	0.0066667
12.6.2012	14.6	67	11	16:30	3.2	0.1066667	6:00	0.4	0.0133333
13.6.2012	2.6	13	6	21:30	1	0.0333333	19:30	0.2	0.0066667
14.6.2012	3.2	16	6	15:00	1.8	0.06	5:30	0.2	0.0066667
15.6.2012	9.6	29	3	13:00	7.6	0.2533333	13:30	0.6	0.02
21.6.2012	16.2	43	4	20:30	8.6	0.2866667	19:30	0.6	0.02
22.6.2012	9	45	7	7:00	2.2	0.0733333	9:00	0.2	0.0066667
25.6.2012	0.4	2	2	11:00	0.2	0.0066667	11:00	0.2	0.0066667
28.6.2012	7.8	16	2	13:30	7.4	0.2466667	7:00	0.4	0.0133333

Table 4.3 Maximum rainfall in 60 min—June 2012

June 2012	Rainfall [mm]	Number of measurements		Max. intensity (60 min)			Min. intensity (60 min)		
		1 min	60 min	Time [day]	Max. rainfall [mm]	Max. intensity [mm/min]	Time [day]	Min. rainfall [mm]	Min. intensity [mm/min]
1.6.2012	0.8	4	3	9:00	0.4	0.0066667	10:00	0.2	0.0033333
4.6.2012	11.4	30	1	0:00	11.4	0.19	0:00	11.4	0.19
5.6.2012	10	50	10	1:00	4.6	0.0766667	2:00	0.2	0.0033333
6.6.2012	0.2	1	1	2:00	0.2	0.0033333	2:00	0.2	0.0033333
7.6.2012	4.2	10	2	13:00	3.8	0.0633333	14:00	0.4	0.0066667
9.6.2012	7.2	29	3	2:00	5.6	0.0933333	22:00	0.2	0.0033333
10.6.2012	5.6	28	2	3:00	3.2	0.0533333	2:00	2.4	0.04
11.6.2012	6.6	20	3	12:00	4.8	0.08	9:00	0.2	0.0033333
12.6.2012	14.6	67	8	17:00	5.8	0.0966667	12:00	0.6	0.01
13.6.2012	2.6	13	5	22:00	1.2	0.02	20:00	0.2	0.0033333
14.6.2012	3.2	16	5	15:00	1.8	0.03	6:00	0.2	0.0033333
15.6.2012	9.6	29	2	13:00	9	0.15	14:00	0.6	0.01
21.6.2012	16.2	43	2	21:00	12.4	0.2066667	20:00	3.8	0.0633333
22.6.2012	9	45	4	7:00	4.2	0.07	9:00	0.2	0.0033333
25.6.2012	0.4	2	2	11:00	0.2	0.0033333	11:00	0.2	0.0033333
28.6.2012	7.8	16	2	14:00	7.4	0.1233333	7:00	0.4	0.0066667

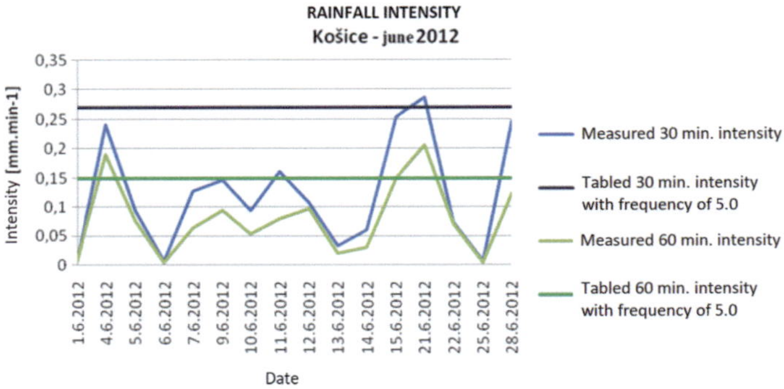

Fig. 4.2 Graph with measured (calculated) and table (SHMU) intensities—June 2012

Two types of study area, located in Košice (Technical University Area) and Prešov (Šariške Lúky), were available for the solution. In both territories, it was possible to define a sufficient number of hydrogeological wells from the archive (GEOFOND—State Geological Institute of Dionýz Štúr (ŠGÚDŠ) in Bratislava), on which hydrodynamic tests were carried out; this provided at least data on the efficiency and reduction of the groundwater level. These correspond to the category II of input data quality for calculating the filtration coefficient.

The following tables present the characteristics of the results thus processed (Tables 4.4, 4.5, 4.6 and 4.7).

Based on Tables 4.4, 4.5, 4.6 and 4.7 and Figs. 4.3 and 4.4, it can be stated that in the evaluated areas, two different collectors were verified quantitatively and qualitatively. Their origin (river sediments) is genetically identical, but sedimentation took place in two different environments with different lithological characteristics. In view of the presented research, for the determination of the value of the filtration coefficient to assess the possibility of rainwater infiltration at an uninterrupted groundwater level, it is important that the valid reliable results correspond to the actual geohydraulic properties of the collectors.

Figure 4.3: The confidence intervals of the calculated mean values and the range of the individual *Y*-values in the quaternary run-offs (*T*—flow coefficient; Y_{min}, Y_{max}—the minimum and maximum values of *Y*; *M*(Y)—the calculated value of the

Table 4.4 Drainage thickness and max. drilling capacity

Rated area	Number of data	Thickness of water layer M [m]		Max. drilling capacity Q [l s^{-1}]	
		Range	Median	Range	Median
Košice—river sediments of Hornád	14	3.0–10.2	5.85	2.0–24.0	10.5
Prešov—river sediments of Sekčov	8	1.7–8.2	3.25	0.15–2.00	0.63

Table 4.5 Values of the flow-rate index Y and permeability index Z from the collectors under investigation

Rated area	Number of data	Flow rate index Y				Index of permeability Z			
		Range	Median	Arithmetic mean	s_Y	Range	Median	Arithmetic mean	s_Z
Hornád	14	6.00–6.94	6.51	6.5	0.07	5.35–6.13	5.79	5.78	0.05
Sekčov	8	4.82–6.03	5.36	5.44	0.16	4.41–5.63	4.75	4.90	0.17

s-standard deviation

Table 4.6 Confidence intervals of the arithmetic mean $M(Y)$ and estimates of average values of the flow coefficient of quaternary aquifers

Evaluated area	Flow rate index Y		Flow rate T [m^2 s^{-1}]		
	$M(Y)$	M_1–M_2	$G(T_1)$–$G(T_2)$	$G(T)$	Class
Hornád	6.50	6.37–6.62	3.36×10^{-3}–6.21×10^{-3}	4.62×10^{-3}	IIa
Sekčov	5.44	5.13–5.75	1.58×10^{-4}–7.29×10^{-4}	3.39×10^{-4}	IIIa

$M(Y)$—the arithmetic mean of the values of the flow index; M_1, M_2—the lower and upper limits of the confidence interval of the arithmetic mean $M(Y)$ for the probability of 90%; the $G(T)$—estimate of the geometric mean of flow-factor T with 90% reliability $G(T_1)$ and $G(T_2)$; class—classification of average flow-rate in terms of the Krásný classification

Table 4.7 Confidence intervals of arithmetic mean $M(Z)$ and estimates of average values of the permeability coefficient of quaternary aquifers

Evaluated area	Flow rate index Z		Flow rate T [m^2 s^{-1}]		
	$M(Z)$	M_1–M_2	$G(k_{f1})$–$G(k_{f2})$	$G(k_f)$	Class
Hornád	5.78	6.68–5.87	6.85×10^{-4}–1.11×10^{-4}	8.81×10^{-4}	IIIa
Sekčov	4.90	4.58–5.22	4.46×10^{-5}–2.15×10^{-4}	9.79×10^{-5}	IVa

$M(Z)$—sample arithmetic mean of values of the permeability index; M_1, M_2—lower and upper limit of confidence interval of arithmetic mean $M(Z)$ for probability 90%; $G(k)$—estimate of the geometric mean of the filtration coefficient k by the limits of the 90% $G(k_1)$ and $G(k_2)$; class—classification of average permeability in terms of the Jetel classification (1982)
In the case of classification of the flow rate and permeability of the evaluated water layers, the symbols characterizing the variability of the heterogeneity of the given hydrogeological environment according to Krásný (1993) were used for each class

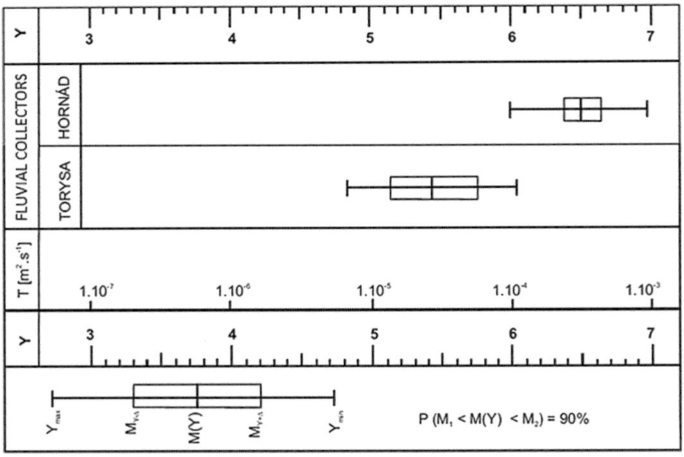

Fig. 4.3 Graphic representation of the flow index in Košice and Prešov

arithmetic mean of the flow index Y; M_1, M_2—the lower and upper limits of the 90% confidence interval of the arithmetic mean of the Y index).

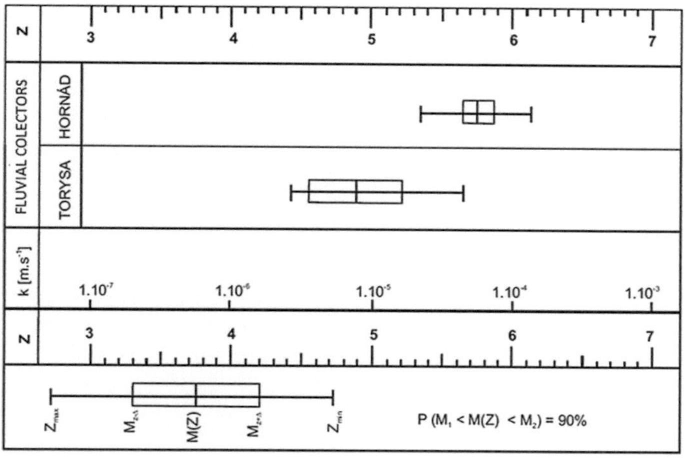

Fig. 4.4 Graphic representation of the permeability index in Košice and Prešov

Figure 4.4: The confidence intervals of the calculated average values and range of the permeability index values *Z* in the quaternary substations (k_f—filtration coefficient; $Z_{\min}$, $Z_{\max}$—minimum and maximum detected transmittance index *Z*; $M(Z)$—calculated selected value of the arithmetic mean of the permeability index *Z*; M_1, M_2—lower and upper limit of the 90% confidence interval of the determination of the arithmetic mean of the permeability index *Z*).

4.2.2 Determination of the Coefficient of Filtration by Laboratory Tests from Grain Curves

The output from sieving and the density test is a grain curve processed by software at the GEO Slovakia Ltd. laboratory. The grain curve (Fig. 4.5) is then used to determine the fraction of the soil and to calculate the coefficient of filtration. These results were compared with the numerical calculations specified in the next chapter.

In order to assess the use of grain curves in determining the values of the filtration coefficient under the conditions in which the precipitation water is to be infiltrated, laboratory analysis of soil samples taken at the Technical University campus was used. The procedure for the grain-size analysis is given in the previous chapter. In order to compare these results, a group of samples was also included in the analysis as part of the older research work carried out in the surroundings of the study area under very similar geological conditions (Baróková 2008, 2009). As a representative layer, which can serve to infiltrate rainwater to the free groundwater level, the selected type of soil was classified in accordance with STN 72 1001 (Classification of soils and rocks from 2010) into Grade G3—gravel with fine-grained soil (symbol GF).

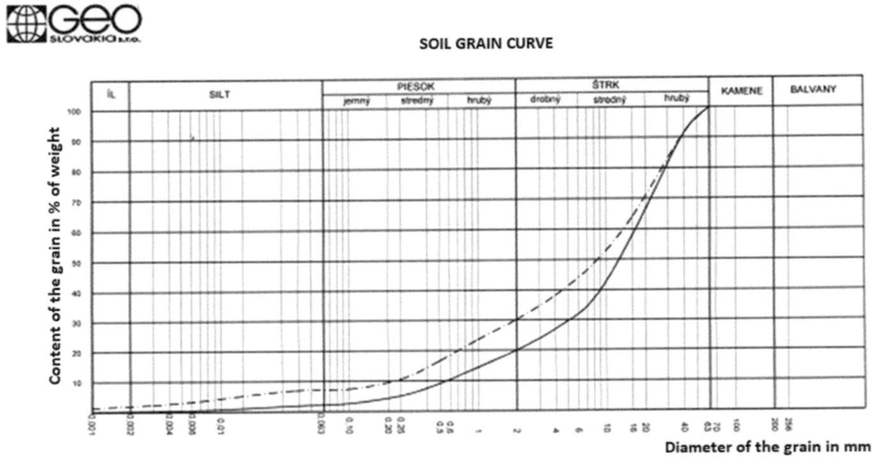

Fig. 4.5 Soil grain curve

The values of the filtration coefficient k_f using the results of the laboratory analyses of gravel with fine-grained soil samples were calculated by three methods generally applied for the indicative determination of this value. The results of these calculations are shown in Table 4.8.

By interpreting the resultant values (Table 4.8) of the filtration coefficient, it can be stated that in any case, they are only indicative values. The different values obtained point this out to the full extent, as sometimes there is a difference of a hundred-fold.

Within the field test, measuring devices were installed to provide the data and values needed for the research.

In the case of grain curves determined from laboratory tests in the framework of this research, k_f ranges from 7.81×10^{-5} to 6.4×10^{-3} m s^{-1}. Compared to other archival works (Baróková 2008, 2009), the differences in values of k_f determined by these methods (Table 4.8) are even more pronounced, from 4.17×10^{-7} to 5.63×10^{-3}m s^{-1}.

It follows from the above that the determination of the value of the filtration coefficient in the laboratory from grain curves does not represent a suitable method for the purpose of rainwater infiltration into the soil due to the very wide range of values.

From the laboratory analysis described in Sect. 2.5, the filtration coefficients were determined for the two model areas described in Sect. 2.1. Grain curves were prepared at GEO Slovakia Ltd. The soil type from a sample from shaft A in Košice was gravel well-grained, and the sample from shaft B was gravel with fine-grained soil (Fig. 4.5). This conclusion was also confirmed by the hydrogeological survey conducted in the area in 2010.

The filtration coefficient is also determined from the grain curve. When calculating the filtration coefficient, the size of gravel grains by falling over 10 and 60% is important (see Sect. 1.4.5.1). The results are summarized in Fig. 4.6.

Table 4.8 Calculation of the filtration coefficient from the grain curves

Grain diameters for gravel with fine soil from the study area by 10, 50, 60 % of the weight								
Hudáková (2013)			Bartóková (2008)			Bartóková (2009)		
10	50	60	10	50	60	10	50	60
d(mm)								
0.5	8.0	16.0	0.2	5.0	9.0	0.1	2.0	4.0
d (m)								
5.0E−04	8.0E−03	1.6E−02	2.0E−04	5.0E−03	9.0E−03	1.0E−04	2.0E−03	4.0E−03
$C_w = d_{60}/d_{10}$								
32.0			45.0			40.0		
d_{10}/d_{10}								
3.1E−02			2.2E−02			2.5E−02		
k according jaky (1994) $k = 100 \cdot d_{50}$ $(d(m))$								
6.4E−03			**2.5E−03**			**4.0E−04**		
k according Hazen (1893) $k = \frac{0.01 \cdot d_{10}^2}{C_w}$								
7.81E−05			**8.89E−06**			**2.50E−06**		
k according Bayer and Schwelger (Bayer 1969) $k = \frac{\gamma}{\eta} \cdot 1.7 \cdot 103\left(\frac{d_{60}}{d_{10}}\right)^{-0.2} \cdot d_{10}^2$								
$(d_{60}/d_{10})^{-0.2}$								
2.00			2.14			2.09		
Pri hodnote $\eta/\gamma = 7.5 \times 10^{-6}$ [m s^{-1}]								
6.57E−04			**1.12E−04**			**2.75E−05**		
Polák (2008)			Spišák (2009)			Durbák (2010)		
10	50	60	10	50	60	10	50	60
–								
0.1	1.5	3.0	0.1	6.0	10.0	0.7	7.5	15.0
–								
5.0E−05	1.5E−03	3.0E−03	1.2E−04	6.0E−03	1.0E−02	7.0E−04	7.5E−03	1.5E−02
–								
60.0			83.3			21.4		
–								
1.7E−02			1.2E−02			4.7E−02		
–								
2.3E−04			**3.6E−03**			**5.6E−03**		
–								
4.17E−07			**1.73E−06**			**2.29E−04**		
–								
–								
2.27			2.42			1.85		
–								
7.45E−06			**4.58E−05**			**1.19E−03**		

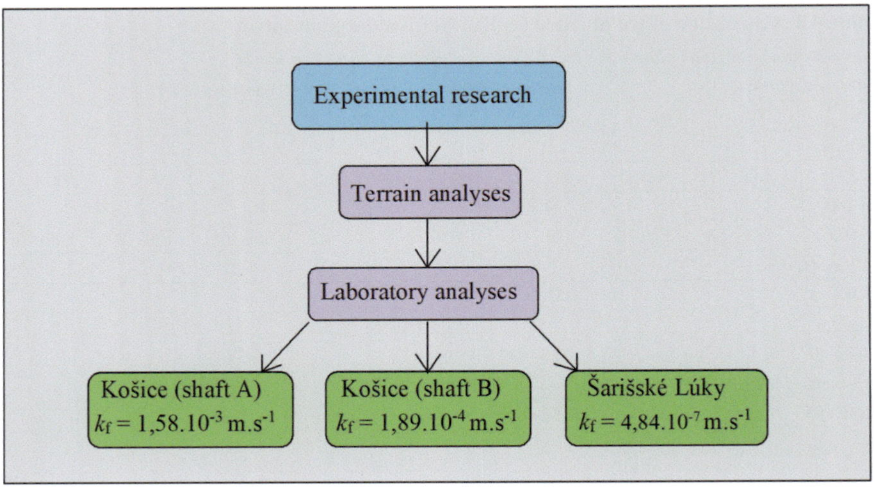

Fig. 4.6 Laboratory results

Figure 4.6 shows that the filtration coefficient for shaft A is 1.58×10^{-3} m s^{-1}, for shaft B in Košice is 1.89×10^{-4} m s^{-1} and for Šarišské Lúky is 4.84×10^{-7} m s^{-1}. This analysis was needed to compare the results of the filtration coefficient determined by numerical analysis.

4.2.3 Calculation of the Filtration Coefficient Using Empirical Relations

Filtration coefficients from the areas of interest in Košice and Prešov were also calculated using empirical formulas according to ČSN 75 9010 and according to the final report from the geological survey of the environment for Sabinov prepared by doc. Ing. Ladislav Tometz, Ph.D., which are recalculated and adapted for the measured values of the research.

The sample calculation according to ČSN 75 9010 is as follows. According to relations 3.4, 3.5, 3.7:

$$Q_I = \frac{0.38}{31260} = 1.22 \times 10^{-5}\ \mathrm{m^3\ s^{-1}}$$
$$A_I = \pi \cdot 0.5^2 = 0.785\ \mathrm{m^2}$$
$$k_f = \frac{2.1.22 \times 10^{-5}}{0.785} = 3 \times 10^{-5}\ \mathrm{m\ s^{-1}}$$

For Košice, the results of the filtration coefficient calculated according to the technical standard for shaft A are shown in Table 4.9 and for shaft B in Table 4.10.

Table 4.9 Calculated filtration coefficient for shaft A in Košice according to ČSN 75 9010

Date of rain event	Time from	Time to	V [m^3]	T [s]	Q_I [m^3 s^{-1}]	A_I [m^2]	k_f [m s^{-1}]
5.11.2013 6.11.2013	13:36	1:00	2.03	41,040	4.95×10^{-5}	0.785	1.26×10^{-4}
9.11.2013 10.11.2013	20:02	5:08	0.77	32,760	2.35×10^{-5}	0.785	5.99×10^{-5}
19.3.2013	2:03	11:18	0.17	33,300	5.11×10^{-6}	0.785	1.30×10^{-5}
14.1.2013 15.1.2013	22:59	12:03	0.12	47,040	2.55×10^{-6}	0.785	6.50×10^{-6}
16.1.2013	0:06	12:39	0.12	45,180	2.66×10^{-6}	0.785	6.76×10^{-6}
19.9.2012 20.9.2012	19:21	15:05	1.25	71,040	1.76×10^{-5}	0.785	4.48×10^{-5}
21.7.2012	0:03	11:02	2.31	39,540	5.84×10^{-5}	0.785	1.49×10^{-4}
16.5.2012 17.5.2012	10:29	1:03	0.3	52,440	5.72×10^{-6}	0.785	1.46×10^{-5}
16.12.2011 17.12.2011	17:08	2:37	0.72	34,140	2.11×10^{-5}	0.785	5.37×10^{-5}
17.12.2011	6:16	15:32	1.16	33,360	3.47×10^{-5}	0.785	8.86×10^{-5}

Table 4.10 Calculated filtration coefficient for shaft B in Košice according to ČSN 75 9010

Date of rain event	Time from	Time to	V [m^3]	T [s]	Q_I [m^3 s^{-1}]	A_I [m^2]	k_f [m s^{-1}]
22.2.2014	1:18	9:59	0.38	31,260	1.22×10^{-5}	0.785	3.10×10^{-5}
9.2.2014 10.2.2014	23:04	6:53	1.14	28,140	4.05×10^{-5}	0.785	1.03×10^{-4}
20.1.2014	3:00	14:17	1.49	40,620	3.67×10^{-5}	0.785	9.35×10^{-5}
5.11.2013 6.11.2013	13:36	1:00	8.61	41,040	2.10×10^{-4}	0.785	5.35×10^{-4}
9.11.2013 10.11.2013	20:02	5:08	4.08	32,760	1.25×10^{-4}	0.785	3.17×10^{-4}
19.3.2013	2:03	11:18	1.87	33,300	5.61×10^{-5}	0.785	1.43×10^{-4}
29.3.2013	7:40	22:55	0.85	54,900	1.54×10^{-5}	0.785	3.95×10^{-5}
6.2.2013 7.2.2013	17:54	3:11	1.32	33,420	3.95×10^{-5}	0.785	1.01×10^{-4}
7.2.2013	5:17	16:57	0.6	42,000	1.43×10^{-5}	0.785	3.64×10^{-5}
14.1.2013 15.1.2013	22:59	12:03	1.55	47,040	3.20×10^{-5}	0.785	8.39×10^{-5}
16.1.2013	0:06	12:39	2.48	45,180	5.49×10^{-5}	0.785	1.40×10^{-4}
21.1.2013	0:23	11:59	0.84	41,760	2.01×10^{-5}	0.785	$5.12.10^{-5}$
21.1.2013 22.1.2013	14:09	0:28	1.03	37,140	2.77×10^{-5}	0.785	7.07×10^{-5}
19.9.2012 20.9.2012	19:21	15:05	7.34	71,040	1.03×10^{-4}	0.785	2.63×10^{-4}
21.7.2012	0:03	11:02	4.93	39,540	1.24×10^{-4}	0.785	3.18×10^{-4}
16.5.2012 17.5.2012	10:29	1:03	1.61	52,440	3.07×10^{-5}	28.26	7.82×10^{-5}

Table 4.11 Calculated filtration coefficient for the shaft in Prešov according to ČSN 75 9010

Date	Time from	Time to	V [m^3]	T [s]	Q_{vsak} [m^3 s^{-1}]	A_{vsak} [m^2]	k_f [m s^{-1}]
2.4.2013 4.4.2013	22:52	3:46	1.72	104,040	1.65×10^{-5}	46.08	7.18×10^{-7}
2.2.2013	12:18	21:50	0.45	34,320	1.31×10^{-5}	46.08	5.69×10^{-7}
15.1.2013 16.1.2013	11:10	14:51	3.15	99,660	3.16×10^{-5}	46.08	1.37×10^{-6}
17.12.2012 18.12.2012	8:05	10:57	0.56	96,720	5.79×10^{-6}	46.08	2.51×10^{-7}
1.11.2012	10:40	23:02	0.12	44,520	2.71×10^{-6}	46.08	1.18×10^{-7}
19.9.2012 20.9.2012	19:49	15:51	3.07	72,120	4.26×10^{-6}	46.08	1.85×10^{-6}

In Prešov, the results of the coefficient of filtration are shown in Table 4.11. The calculation is similar to that in Košice, as it is also according to ČSN 75 9010. According to relations 60, 62, 63,

$$Q_I = \frac{1.72}{104040} = 1.65 \times 10^{-5}\ \text{m}^3\ \text{s}^{-1}$$

$$A_I = 6.4.7.2 = 46.08\ \text{m}^2$$

$$k_f = \frac{2.1.65.10^{-5}}{46.08} = 7 \times 10^{-7}\ \text{m}\ \text{s}^{-1}$$

Within the numerical analysis to calculate the filtration coefficient in Košice, two formulas were used. The first formula is according to the cited standard ČSN 75 9010 and the second according to the final report of assoc. prof. Ladislav Tometz from the geological survey of the Sabinov environment, which differs in the input values. The comparison of the results of the mentioned empirical equations can be seen in Figs. 4.7 and 4.8.

From the comparison of the filtration coefficients that were calculated using both approaches and the comparison with the laboratory coefficients of filtration ($k_f = 1.58 \times 10^{-3}$ m s^{-1} and $k_f = 1.89 \times 10^{-4}$ m s^{-1}), it follows that for the given model area, approximately, the same filtration coefficient is stated, as confirmed by the DWA-A 138E table (Fig. 2.5). The values range from 10^{-4} to 10^{-6}, which converts coarse sand through dusty sand to clay dust.

In the conditions of model territory II in Šarišské Lúky, the results of the coefficient of filtration according to the empirical formula ČSN 75 9010 are shown in Fig. 4.9. For comparison, the result of the laboratory coefficient of filtration is $k_f = 4.84 \times 10^{-7}$ m s^{-1}.

For calculation of the filtration coefficient according to the final report, it was not possible to obtain any relevant data, since the measurement started in September 2012 and ended in August 2013.

Fig. 4.7 Summary of the numerical analysis for the filtration coefficient calculations from the measured values of infiltration shaft A in Košice

Infiltration shaft A Košice	
ČSN 759010 k_f [m.s^{-1}]	Geoton Ltd. k_f [m.s^{-1}]
$1{,}26.10^{-4}$	$1{,}04.10^{-6}$
$5{,}99.10^{-5}$	$3{,}12.10^{-6}$
$1{,}30.10^{-5}$	$1{,}04.10^{-6}$
$6{,}50.10^{-6}$	$2{,}08.10^{-6}$
$6{,}76.10^{-6}$	$2{,}08.10^{-6}$
$4{,}48.10^{-5}$	$7{,}20.10^{-6}$
$1{,}49.10^{-4}$	$6{,}00.10^{-6}$
$1{,}46.10^{-5}$	$1{,}20.10^{-6}$
$5{,}37.10^{-5}$	$2{,}40.10^{-6}$
$8{,}86.10^{-5}$	$1{,}20.10^{-6}$

Fig. 4.8 Summary of the numerical analysis for the filtration coefficient calculations from the measured values of infiltration shaft B in Košíce

Infiltration shaft B Košice	
ČSN 759010 k_f [m.s^{-1}]	Geoton Ltd. k_f [m.s^{-1}]
$3{,}10.10^{-5}$	$1{,}04.10^{-6}$
$1{,}03.10^{-4}$	$1{,}04.10^{-6}$
$9{,}35.10^{-5}$	$1{,}04.10^{-6}$
$5{,}35.10^{-4}$	$1{,}04.10^{-6}$
$3{,}17.10^{-4}$	$3{,}12.10^{-6}$
$1{,}43.10^{-4}$	$1{,}04.10^{-6}$
$3{,}95.10^{-5}$	$2{,}08.10^{-6}$
$1{,}01.10^{-4}$	$1{,}04.10^{-6}$
$3{,}64.10^{-5}$	$1{,}04.10^{-6}$
$8{,}39.10^{-5}$	$2{,}08.10^{-6}$
$1{,}40.10^{-4}$	$2{,}08.10^{-6}$
$5{,}12.10^{-5}$	$2{,}08.10^{-6}$
$7{,}07.10^{-5}$	$1{,}04.10^{-6}$
$2{,}63.10^{-4}$	$7{,}20.10^{-6}$
$3{,}18.10^{-4}$	$6{,}00.10^{-6}$
$7{,}82.10^{-5}$	$1{,}20.10^{-6}$

Fig. 4.9 Summary of the numerical analysis for filtration coefficient calculations from the measured values of the infiltration blocks in Šarišské Lúky

Infiltration gallery Šarišské Lúky
ČSN 759010 k_f **[m.s^-1]**
7,18.10^-7
5,69.10^-7
1,37.10^-6
2,51.10^-7
1,18.10^-7
1,85.10^-6

The measurement was terminated due to the malfunctioning and clogging of the infiltration device. One of the reasons for this was the incorrect design of the infiltration device based on bad input data. The laboratory filtration coefficient was stated as 4.84×10^{-7} m s^{-1}, and a filtration coefficient value of 8.2×10^{-5} m s^{-1} was calculated for the design of the infiltration gallery. The whole research facility (Fig. 4.10) was clogged not only due to poor estimation or determination of the filtration coefficient but also due to the fact that the filtration shaft did not function and did not collect sediments. For this reason, it is possible to recommend from the alternative solutions for infiltration devices the infiltration shafts used in model area I, which were designed correctly.

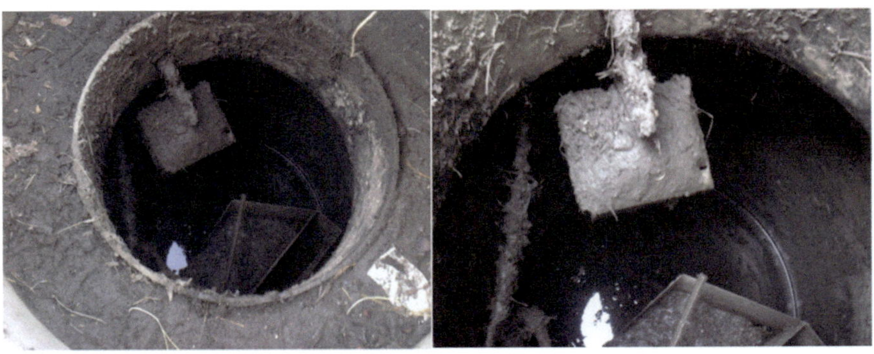

Fig. 4.10 Clogged shaft in model area II

It follows that the functionality of the infiltration equipment depends mainly on the design.

4.3 Verification of the Mathematical–Physical Model for Determination of the Time

For the modelling of the emptying time, it was necessary to know the parameters that affect infiltration, in order to calculate the dimensionless arguments and the regression coefficients. The values of the dimensionless arguments π_1 and π_2 were calculated from the measured data (Table 4.12).

The difference between the measured and calculated values of the emptying time was determined as a relative error (uncertainty).

The dependence of the dimensionless arguments π_1 and π_2 (from Table 4.12) is presented in Fig. 4.11 and can be expressed in the form:

$$\pi_1 = \mathrm{A} \cdot \pi_2^{\mathrm{B}} \tag{4.1}$$

$$y = 1.0947 \cdot x^{-0.374}$$

where: y—dependent variable π_1, x—independent variable π_2.

The values of the location coefficient and the regression coefficient are

$$A = 1.0947,$$
$$B = -0.374.$$

The obtained location coefficient A and regression coefficient B were used to verify the model—i.e., determining the length of time by Eq. (3.20).

In Fig. 4.12, the measured and calculated values of the emptying time are compared. High to extremely high deviations occur particularly at times of high flow rates. A mathematical–physical model response to such situations is not possible.

Uncertainty—the average error of a given model is calculated from the equation:

$$\sigma = \frac{1}{n} . \sum_{i-n}^{n} \frac{\left| T_{nam} - T_{vyp} \right|}{T_{nam}} .100 \tag{4.2}$$

It was also necessary to determine the significance of uncertainty (Table 4.13). For natural processes, we recommend using the following values:

From the results of determining the emptying time, the significance of uncertainty can be characterized as low to medium.

Table 4.12 Values of relevant parameters, dimensionless arguments, measured and calculated values of emptying time, uncertainty

No.	Filtration coefficient	Infiltration area	Rain intensity	Emptying time			Emptying time (measured)	Emptying time (calculated)	Uncertainty
	k_f	S	i	T_{vs}	π_2	π_1	T_{nam}	T_{vyp}	σ
	[m s^{-1}]	[m^2]	[m s^{-1}]	[s]			[s]	[s]	[%]
1	1.58E−03	0.785	7.331E−07	5580	0.0004627	9.95	5580	10,845.930	94.37
2	1.58E−03	0.785	4.70E−07	9360	0.0002975	16.69	9360	12,793.120	36.68
3	1.58E−03	0.785	4.32E−07	27,840	0.0002737	49.65	27840	13,198.883	52.59
4	1.58E−03	0.785	4.62E−07	26,880	0.0002922	47.93	26,880	12,879.408	52.09
5	1.58E−03	0.785	4.20E−07	13,800	0.0002657	24.61	13,800	13,345.717	3.29
6	1.58E−03	0.785	2.10E−07	12,660	0.0001328	22.58	12,660	17,299.900	36.65
7	1.89E−04	0.785	7.31E−07	59,580	0.0038677	12.71	59580	40,978.972	31.22
8	1.89E−04	0.785	4.70E−07	55,500	0.0024872	11.84	55,500	48,336.002	12.91
9	1.89E−04	0.785	4.32E−07	57,780	0.002288	12.33	57,780	49,869.088	13.69
10	1.89E−04	0.785	4.62E−07	31,800	0.0024429	6.78	31,800	48,662.021	53.03
11	1.89E−04	0.785	4.20E−07	56,700	0.0022213	12.10	56,700	50,423.867	11.07
12	1.89E−04	0.785	2.10E−07	35,820	0.0011099	7.64	35,820	65,363.882	82.48
									40.01

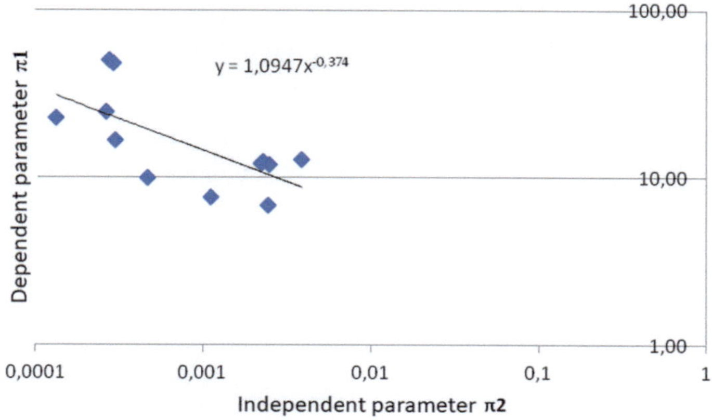

Fig. 4.11 Dimensionless arguments and a regression line displayed in logarithmic coordinates

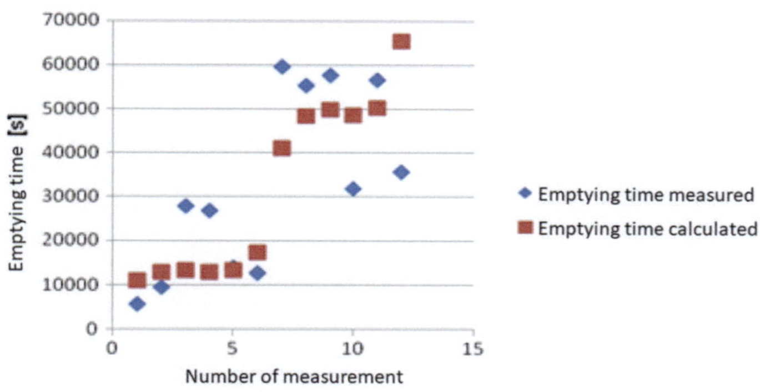

Fig. 4.12 Measured and calculated emptying time

Table 4.13 Significance of uncertainty

Uncertainty—σ [%]	Significance
0–20	Insignificant
20–40	Low
40–60	Medium
60–80	High
80 and more	Undesirable

References

Baróková M (2008) Košice – areál Univerzity veterinárneho lekárstva – novostavba – Chirurgia a PGA malých zvierat, podrobný IGP. Manuskript – achív ŠGÚDŠ Bratislava, 29 s

Baróková M (2009) Košice – Drevný trh – Bytový dom, podrobný IGP. Manuskript – achív ŠGÚDŠ Bratislava, 44 s

Hudáková G (2013) Research of rainwater infiltration in the Košice region. Colloquium on Landscape Management. Brno: Mendel University in Brno, pp. 19–22

Chapter 5
Evaluation of Achieved Results

Abstract When constructing any building, it is necessary to take into account all aspects of the functionality and operational ability of the building. One important task is the drainage of rainwater from the roofs, car parks, roads or pavements belonging to the buildings. A very advantageous and simple solution has recently been to drain this water into the sewer system, but with the continual development of cities, the construction of new buildings, existing sewerage pipes and lines becomes under-dimensioned, and their constant adaptation to actual needs is impossible. The current state of sewerage systems is insufficient in this respect, and so in the case of torrential rains, if there is more rain in a few minutes than during the whole month, there will be local floods.

5.1 Relation to Legislation

The governments of many countries (Germany, Switzerland, Great Britain, the Netherlands, USA, Canada, Australia and others) have been dealing with this problem for some time. At present, new European directives are already addressing this issue. They instruct you to manage the rainwater by infiltration, directly at the impact site, if possible. If not, drainage of rainwater into the sewerage or the watercourse is only possible through a retention reservoir.

In particular, this is related to EU Directive 271/91/EEC, which regulates the proposal of infiltration systems, following the Report on Water Scarcity and Droughts in the European Union and others.

In the territory of Slovakia, infiltration/percolation systems are also used which, with the help of filters and/or oil separators, can drain rainwater, though not only from the car parks and pavements. Of course, certain procedures, technical standards and regulations are used to design proper infiltration facilities.

There is currently no suitable technical standard in Slovakia and no regulation that is devoted to the design of an infiltration facility itself. There are several laws relating to this issue. Law No. 364/2004 Coll. on Water, as amended, refers to the management of rainwater, but it is inapplicable for the design of infiltration systems. Decree of the Ministry of the Environment No. 532/2002 Coll. deals with precipitation water.

M. Zeleňáková et al., *Rainwater Infiltration in Urban Areas*,
Water Science and Technology Library 89,
https://doi.org/10.1007/978-3-030-34698-0_5

Specifically, it is § 9—connection of a building to the local distribution of technical equipment of the territory, par. 3 states: "The building must preferably be connected to a public sewerage system if it has sufficient capacity or if waste water disposal facilities are to be implemented. If the discharged waste water does not meet the conditions for discharging into the public sewer system, it is necessary to design and set up a pre-treatment facility."

Further, there are Law No. 442/2002 Coll. on Public Water Supply and Sewerage, as amended, and Decree No. 269/2010 Coll. as amended—on achieving good water status and more legislative. None of them, however, implements the design and sizing of infiltration devices.

The infiltration devices should be located in such a way that they do not cause any damage either to the drained buildings or to neighboring buildings and other facilities. Designers in the design of infiltration devices adhere to minimum distances from objects as well as from groundwater level in accordance with German technical standard DWA-A 138E or Czech technical standard ČSN 75 9010.

Czech technical standard ČSN 75 9010 describes the scope and method of carrying out a geological survey, determines the limiting conditions for the absorption of precipitation waters, describes the procedure and examples of calculations of retention volumes of the infiltration equipment and deals with the degree of safety against overfilling and overflowing of the infiltration equipment. At the same time, the standard provides a basic overview of the currently used surface and underground infiltration devices, and the annexes to the standard contain updated tables of rainfall design intensities in the Czech Republic.

The German standard DWA-A 138E defines in detail the infiltration process, specifies the qualitative and quantitative principles of the design of the infiltration equipment, characterizes the individual components of the system, describes the phases in the process of designing and installing the infiltration equipment and finally states the economic effect of their use.

The basic differences in these standards are shown in Table 5.1.

In Slovakia, the German standard DWA-A 138 and Czech standard ČSN 75 9010, which was created on the basis of the mentioned German standard, are used. But the fact that the appropriate standard for the design of infiltration devices is not created for a specific country is a problem. Therefore, a certain unification of the computational and design procedures for the specific territory would be necessary, since both standards contain certain specifications relating to the country in question. Some designers are governed by a Czech, others by a German standard, with places in which there are adequate differences in procedures and thus in the results of the solution to the problem.

5.2 Research Results

The tables and the graphs were worked out from the field measurements of the precipitation totals in the study areas. The precipitation totals in Košice did not exceed the totals with a periodicity of 0.2, which according to the German standard DWA-A

Table 5.1 Comparison of ČSN 75 9010 and DWA-A 138E

<table>
<tr><th>Czech standard ČSN 75 9010</th><th>German standard DWA-A 138E</th></tr>
<tr><td colspan="2">(A) Placement of the level of the bottom of the infiltration device with respect to the maximum groundwater level</td></tr>
<tr><td>The level of bottom of the infiltration device should be at least 1.0 m above the maximum groundwater level. In exceptional cases, this distance can be reduced based on an hydrogeological survey</td><td>The level of the bottom of the infiltration device should be at least 1.5 m above the maximum groundwater level</td></tr>
<tr><td colspan="2">(B) The distance of the infiltration device from the building</td></tr>
<tr><td>The distance of the infiltration device from the building should ensure a maximum groundwater level that will not endanger underground spaces of the building. The distance X(m) of the infiltration device from the building whose underground level is below the maximum level of water in the infiltration device (Fig. 5.1) is determined according to the empirical relation:
$X = X_1 + X_2$ (5.1)
for the distance X_1, the relationship is:
$X_1 = \frac{h+0,5}{15 \cdot k_v^{0,25}} + 2$ (5.2)
h—the height difference between the maximum water level in the infiltration device Y_1 and the level of the underground floor Y_2 (m). If the maximum water level in the infiltration device is lower than the level of the underground floor, we set as $h = 0$ m.
X_2 is the extension of the trench bottom (m). If its value is unknown, it is assumed to be $X_2 = 2$ m. However, the infiltration device should not be in the backup for the foundations of the building</td><td>If groundwater level is always below the bottom edge of the foundation, there is no need to isolate it against moisture; the distance of the infiltration device from the building excavation at the foundation should not be less than 1.5 times the depth of the trench/dump of the building pit (Figure)
In cases where this is not available, for example, for old objects, the depth of the base plus 0.5 m can be used
For non-waterproofing buildings, it is advisable not to install infiltration systems in the immediate vicinity of buildings, e.g., at the excavation site</td></tr>
<tr><td colspan="2">Note—The Czech standard addresses the distance of the infiltration facility from the object according to the formula (1)
When solving the same infiltration facilities by the two standards, relatively large differences could also arise in some cases</td></tr>
<tr><td>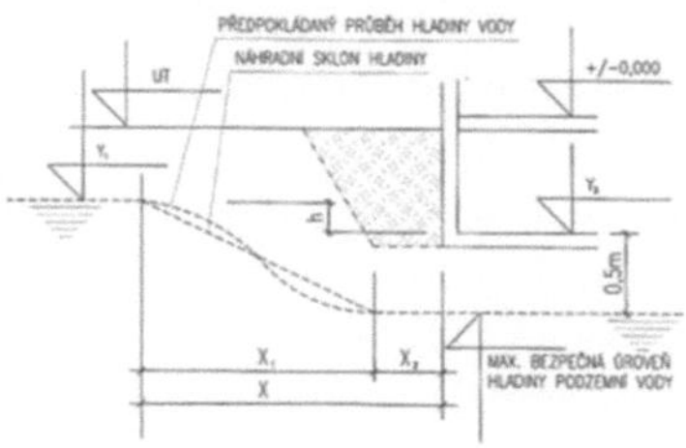</td><td>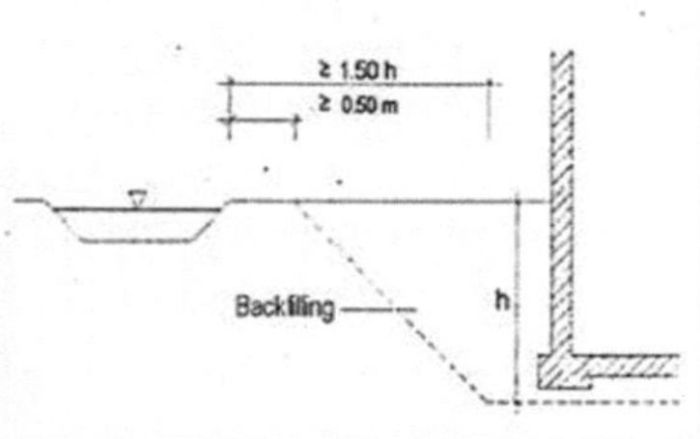</td></tr>
<tr><td colspan="2">Figure determining the distance of the infiltration device from the building</td></tr>
</table>

(continued)

Table 5.1 (continued)

Czech standard ČSN 75 9010	German standard DWA-A 138E
(C) Calculation of the retention volume and the emptying time of the infiltration device	
It considers the coefficient of hydraulic conductivity (coefficient of infiltration) (k_v)	It considers the coefficient of filtration (k_f)
Both variables can be determined both as terrain tests and calculations using formulas, but it is very important that they do not confuse them in the calculations	
(D) Maximum emptying time of infiltration facility	
72 h	24 h
(E) Basis for the development of the technical standard	
ČSN 75 9010 does not use as the basis ČSN 75 6261—rainwater tanks valid in the Czech Republic in the previous period	The standard has been revised several times since its publication in 1990, thanks to the experience and facts found in practice The German standard DWA-A 138 takes some data on designing the retention volume of the rain tank from the previous standard DWA-A 117 dimensioning of rainwater tanks

138 and Czech standard ČSN 75 9010 is used for the design of infiltration equipment. From the evaluated rainfall events, the maximum and minimum precipitation intensity for Košice were calculated and compared with the precipitation rainfall tables according to SHMI for the Košice precipitation station. Only intensity with a periodicity of 5 was exceeded. Rain intensity values over the monitored three-year period did not exceed intensity with a periodicity of 0.2; 0.5; 1; 2 in the monitored station in Košice. It follows that the use of the rain periodicity of 0.2 for the sizing of infiltration equipment is appropriate. The duration of the precipitation compartments processed by the Slovak Hydrometeorological Institute for the precipitation station in Slovakia is still applicable, despite the old processing date of the tables (year 1968). The validity of these tables has been confirmed, and their usability is also warranted at present, in the period when climatic changes are occurring.

One of the main objectives of employing infiltration devices is efficiency, the slow and gradual emptying of containers, which should not exceed 72 h (3 days) according to ČSN 75 9010 and 24 h (1 day) according to DWA-A 138. This was confirmed for each observed rainfall event. Within the presented research, a model for the calculation of the emptying time that can be used in the variation of the area of the infiltration device was developed.

This requirement is closely related to soil parameters and hydrogeological conditions in the subsoil. According to available regulations, the bottom of the infiltration facility should be located at least 1.0 m, and in the case of infiltration shafts at least 1.5 m, above the maximum groundwater level (STN 75 6402), and the permeability of the infiltration layer should be in the range of $k_f = 5.10^{-3}$–5.10^{-6} m s^{-1}.

The bottom of the infiltration shaft located in the TU Košice campus is located in a layer of gravel with a fraction of fine-grained soils, whose $k_f = 1.10^{-3}$–1.10^{-4} m s^{-1}. These are neo-geological sediments that create favorable conditions for concentrating a huge amount of groundwater. This is the reason why in the spring months, water is kept at the bottom of the shaft flows very slowly. It is associated with a high level of groundwater that is higher in the spring months after the snow melts, as well as with the placement of the bottom of the shaft as well as with the geoengineering profile, where beneath a gravel layer with a fine grain soil (symbol GF, class G3) is a clay gravel (GC, class G5) whose filtration capability is lower.

Figures 5.1 and 5.2 present a summary of the results of the filtration coefficients determined by various methods.

The hydrodynamic test in some cases coincides with ČSN 75 9010, the final report by Geton Ltd., as well as with laboratory analyses.

In conclusion, it is necessary to state that the basis of an optimally designed infiltration system is, of course, a good geological and hydrogeological survey. Determination of the coefficient of filtration, the coefficient of hydraulic conductivity of the first layer in contact with the infiltration device, but also the flow properties of other soil layers are all important. The fact that these characteristics are precisely determined is that they are crucially involved in calculating the emptying time of the infiltration device, thereby contributing to the optimal design of the device itself. In the case of inaccurate data, the risk of overfilling of the shaft is likely to occur at a time of extreme collision (as was presented for Šarišské Lúky study area).

The problem is the fact that Slovakia as well as other mainly Central, East and South European countries lack valid legislation that precisely defines the input requirements for the design of infiltration facilities. New legislation should therefore address not only the determination of the conditions for the accumulation of rainwater but also the scope and method of carrying out geological and hydrogeological surveys.

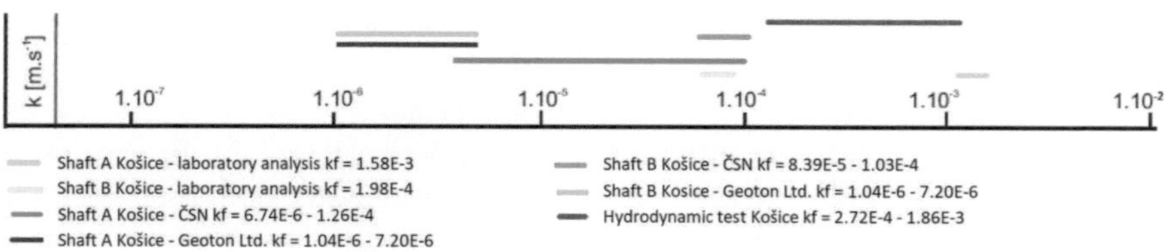

Fig. 5.1 Diagram of results of filtration coefficients in Košice determined by different methods

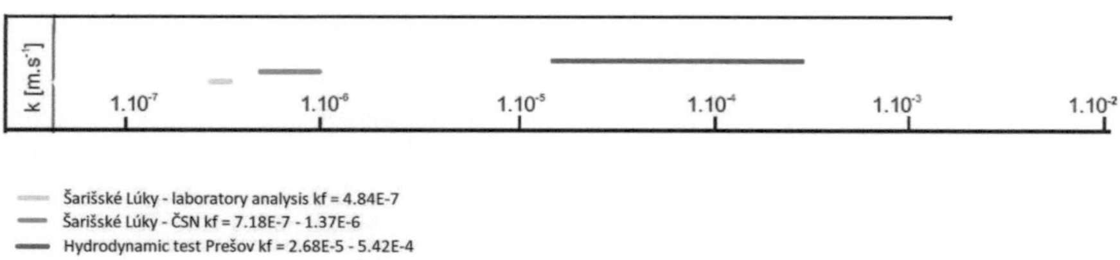

Fig. 5.2 Diagram of results of filtration coefficients in Prešov determined by different methods

5.3 Design of Infiltration Devices

A geological survey is one of the most important bases for the design of infiltration equipment. It should be aimed at acquiring the necessary knowledge about the hydrogeological, engineering and geotechnical conditions of the given site.

Important background for the geological survey itself is

- maps of the appropriate scale corresponding to the relevant stage of the construction documentation, including the plan and elevation, and in the case of line constructions, also longitudinal sections;
- basic data of the building (method and basement level of the building, ground plan of the drainage area as well as the envisaged location of the infiltration device).

A geological survey (indicative, detailed or supplementary) must always be carried out with regard to any decision taken about the accumulation of precipitation water in a certain locality. Its scope is dependent on the character and type of the construction. The survey should be focused on detailed engineering and hydrogeological information. It is very limiting if the survey is focused only on data from maps and hydrological data.

The characteristic of soil permeability is a significant parameter for the design of the infiltration equipment. The layer into which the infiltration device is placed should be suitable for infiltration; it should have suitable porosity and permeability characterized by a coefficient of filtration k_f in the range of $k_f = 10^{-3}$–10^{-6} m s^{-1}. At such values, the k_f assumes that there are no problems in practice. At $k_f \geq 10^{-3}$ m s^{-1}, infiltration takes place very quickly. Assuming that the value of $k_f < 10^{-6}$ m s^{-1}, the infiltration would likely take place very slowly, and draining the precipitation water by the infiltration device would be poorly efficient. The low coefficient of filtration of the soil layer is closely related to the emptying time of the infiltration device, as this should be completely empty before the inflow of new rainfall.

Designing an infiltration device is not a simple matter. First of all, it is necessary to decide whether we want the infiltration device to be used only for the absorption of rainwater into the soil, or to keep the rainwater in some way, or to keep it in case of the first wave of water during heavy rains. The qualitative aspect of the water is also not negligible, because from the point of view of nature conservation, we must know the quality of water infiltrating into the soil. With infiltration of precipitation water, the values of the indicators of permissible groundwater pollution should not be exceeded. Only precipitation waters with permissible pollution indicator values can be infiltrated into the rock environment without previous measures.

In addition, other factors are important, such as the nature and size of the drained area, the annual rainfall in a given location, the properties of the soil where the storage facility will be located, the groundwater level, the possibility of overflowing of the infiltration device, the importance and character of the building and its surroundings, the consequences of the flooding of the building, etc.

For the design and eventual assessment of an infiltration facility by means of a long-term simulation of rainfall ratios in the area of interest, valid historical rainfall data with a minimum length of 10 years for a design for rainfall with periodicity of $p = 0.2$ year^{-1} and data for 20 years for a design for rainfall with periodicity of $p = 0.1$ year^{-1}.

Validation of precipitation data means thorough control, homogenization and transmission of primary rainfall events into historical rainfall events and subsequent statistical evaluation.

If climate analysis is to be meaningful, it must be based on quality data. In the case of precipitation totals over a short period of time, the temporal and spatial variability of these values is very high. For a long-time series of rainfall intensities when these stations are very distant from each other, the correlation between the time series is very low. Validation of rainfall data (primary data obtained from an automatic precipitation station or ombrographic digitization from the SHMI database) therefore includes a sequence of statistical tests to evaluate non-homogeneity in the primary record, adding outages using reference ranges.

Meteorological stations measure rainfall over the long-term, and the results of the observations are given in the form of tables for various probable rainfall intensities, 1 × per year, 1 × every 5 years, 1 × every 10 years. The table for a given location should be available, and if there is no climate station at a given location, the tables must be from the stations closest to the site.

Determination of proposed precipitation water quantities for collection in a manner other than taking data from the SHMI tables is to be documented in the technical report or in the hydrotechnical calculations by publishing technical procedures, which are the validation of the precipitation data.

For the design of a proper rainwater drainage system, the most important criterion is the quality of precipitation surface waters and the suitability of the rock environment for soaking. Designing, dimensioning and construction is certainly equally crucial:

- diagnosis of broader relationships for assessing the impact of groundwater on an object;
- terrain configuration;
- the direction of groundwater flow;
- the depth of the groundwater level from the terrain;
- the calculated hydraulic gradient at the location of the object $I = (h_1 - h_2)/\Delta L$;
- airspace of the recipient (watercourse).

Groundwater in the space between the surface area and the impermeable subsoil can be divided into two zones, the upper zone being the aeration zone (the aerated soil band), and below it is the saturated zone filled with groundwater. In the aeration zone, after laying the well, permanent vegetation from the point of view of the moisture ensures the soil water, regardless of the distance at which it is located. As an example, I would mention the installation of two shafts (the TUKE campus) from the PK6 building, where they were realized at a distance of 2600 mm and 2000 mm from an existing permanent pine tree. The sink well at the PK5 building (Fig. 5.3) was at a distance of 2000 mm. Since 2006 (the year of implementation of the tapping facility), the plant has not shown any changes that would indicate a lack of moisture.

Fig. 5.3 Injection shaft at building PK5—premises of the Technical University in Košice

Correct sizing of the intake shaft is one of the most important parts of its design. Induction devices not only help to break down the sewer networks, but by allowing rainwater to enter the ground as it is under natural conditions, it helps to local natural regeneration of groundwater and beneficial effects on chemical and biological conditions above and below the surface of the terrain.

The determination of the principles and method of designing drinking wells for professional practice is of benefit to Slovakia. It is also important to use proven experience, both in terms of design, implementation and operation of infiltration equipment.

Chapter 6
Overview of Rainwater Management Facilities

Abstract Implementation of European legislation into our system of technical standards introduces a new philosophy of water drainage and protection of watercourses. The aim of this new approach to town drainage is regulation, reducing rainfall flows, reducing the frequency of floods, reducing overall transposed pollution and minimizing its impact on the recipient, and preserving watercourses and groundwater resources in good condition to be able to fulfill their additional functions and to achieve sustainable urban development.

6.1 Surface Infiltration of Rainwater

One of the easiest ways to manage rainwater in urbanized areas is to drain them to the ground. For this purpose, various technical solutions of devices enabling direct contact of rainwater with the ground are used (Fig. 6.1). In some of them, it is also possible to carry out physical processes of sewage pretreatment, such as filtration and sedimentation as well as biological and chemical processes.

In the case of small amounts of rainwater, mainly from residential areas to their infiltration, infiltration surfaces can be used with the top layer of soil sown with grass. These types of surfaces are not suitable for communication.

In the case of areas intended for pedestrian traffic or cars, surfaces with increased durability may be used. They are made of loose materials, such as gravel, sandstone and broken stone, as well as from various cladding, most often prefabricated. Prefabricated panels have the appropriate surface perforation, which can be filled with filtering material (Fig. 6.2) or soil and sown with a mixture of grasses (Fig. 6.3). Thanks to this, such solutions also have the potential to purify rainwater in filtration, sedimentation and biochemical processes. In this way, favorable landscape and environmental effects are obtained.

The use of perforated surfaces instead of a surface tightly separating the soil from the inflow of rainwater gives the possibility of a significant reduction in the size of rainwater runoff from the area. This size can be up to several dozen percent depending on the cladding used, the number of perforations, the material filling the holes, the slope of the area, the intensity of rain and its duration.

M. Zeleňáková et al., *Rainwater Infiltration in Urban Areas*,
Water Science and Technology Library 89,
https://doi.org/10.1007/978-3-030-34698-0_6

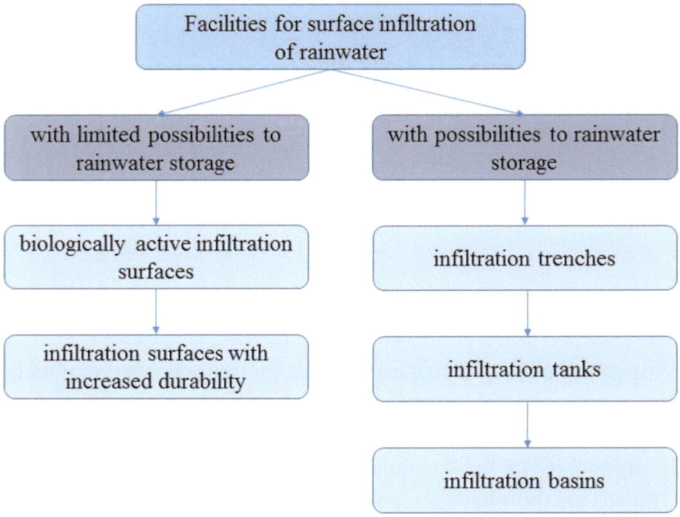

Fig. 6.1 Types of devices for surface rainwater infiltration

Fig. 6.2 Example of a perforated plate with gravel filling (photo Daniel Słyś)

Infiltration trenches are devices that are used to infiltrate rainwater flowing from the surface of roads with low intensity of road traffic and low pollution. They are made as longitudinal depressions of a terrain with a triangular or trapezoidal cross section. They are located along roads, pavements and parking lots. Slopes of infiltration trench are made of humus and sown with a mixture of grasses. The top layer of the bottom of the trench is made of peat mixed with sand and sown with grass under which there is an absorbent layer with a high filtration coefficient k_f.

Fig. 6.3 Example of a paved surface sown with grass (photo Daniel Słyś)

The infiltration basin is a device in the form of a field cavity, whose task is short-term retention and drainage of rainwater to the ground. The basin can be used in soils characterized by at least average values of filtration coefficient k_f. The surface of the basin should be sown with grass, whose task is soil stabilization, intensification of the rainwater purification process and creation of a device with favorable esthetic features. The time of retention of rainwater in the basin should not usually exceed 1–2 days, due to the possibility of dying the vegetation coverage of the basin and clogging of the soil pores in the device. Depending on the amount of rainwater supplied to the infiltration basin, individual devices (Fig. 6.4) or systems of cooperating basins (Fig. 6.5) are used (Fig. 6.6).

Infiltration basins are devices with low filling and low capacity. In the case of larger inflows of rainwater, infiltration tanks are used in the form of properly shaped, open Earth tanks made in permeable soils, which are used to soak rainwater. Similar to infiltration basins, sedimentation and filtration processes as well as biological and biochemical transformation of pollutants take place during the process of retaining rainwater and their infiltration into the soil in infiltration tanks. For esthetic reasons, it is recommended to shape the banks and the bottom of the tank imitating the natural terrain (Fig. 6.7).

Properly designed and integrated into the surroundings, the infiltration tank can be an effective element of the landscape.

Fig. 6.4 Example of an infiltration trench (photo Daniel Słyś)

6.2 Underground Infiltration of Rainwater

Underground storm water infiltration due to its long-lasting character is carried out in devices with specific retention capacity. The classification of these solutions is shown in Fig. 6.8. Depending on the size of the drainage surface, underground infiltration can be carried out in individual devices or their groups. These devices improve the quality of drained rainwater as a result of sedimentation and filtration processes as well as a whole range of biological, chemical and physical processes.

Infiltration boxes are underground devices made of polypropylene truss plates (Fig. 6.9). Due to their considerable capacity, they can fulfill both retention and infiltration role. Their basic task is to create conditions for soaking rainwater into the ground. Polypropylene, from which the boxes are made, provides chemical neutrality to rainwater and groundwater. The construction of the boxes means that despite considerable perforation and lightness of the structure, they are characterized by high mechanical strength. The construction of drainage boxes is characterized by a high proportion of openings outflow which depending on the box producent ranges from 40 to 75%. If there is a need to discharge large amounts of rainwater into the

Fig. 6.5 Example of the implementation of a single infiltration basin (photo Daniel Słyś)

Fig. 6.6 Example of the implementation of the infiltration basin complex in the housing estate (photo Daniel Słyś)

Fig. 6.7 Example of the infiltration tank (photo Daniel Słyś)

Fig. 6.8 Types of devices for underground rainwater infiltration

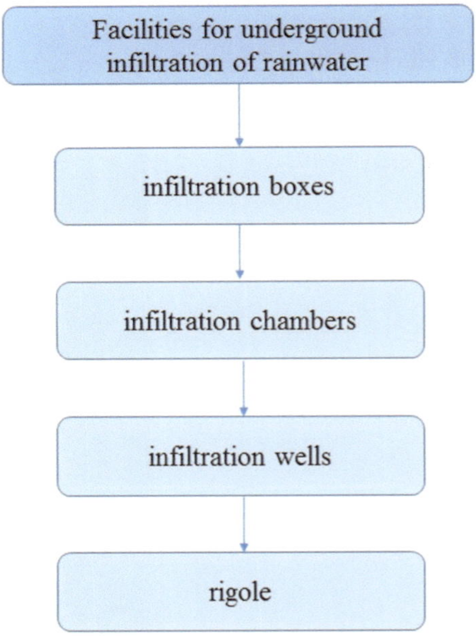

Fig. 6.9 Example of the implementation of the Wavin infiltration boxes

ground, the drainage boxes are combined into one large arrangement of boxes. The retention capacity of a single box is usually around 200–500 dm^3.

A similar function is that the drainage boxes fill the infiltration chambers. These devices, made in the form of plastic arches with an open bottom and perforated walls, have significant retention capacity and high efficiency in draining water to the ground. Single chambers have a capacity of 0.25–0.50 m^3. The modular construction of the chambers, just like the boxes, enables the creation of extensive drainage facilities with unlimited capacity (Fig. 6.10). The low height of the modules is also an advantage, which is particularly important in areas with a high level of groundwater.

The boxes and drainage chambers should be preceded by rainwater cleaning devices. Depending on the method of management of the drained area and the type of contamination flowing with rainwater, settling tanks, filters and separators of petroleum substances may be used.

Due to the possibility of groundwater contamination with substances contained in rainwater, it is recommended to place boxes and drainage chambers at least 1 m above the groundwater level in the case of "clean" surface dewatering. In the case of drainage of heavily contaminated surfaces, this distance should be significantly increased, or these waters should not be directly discharged to underground infiltration equipment. It is then recommended to use surface infiltration devices that provide greater safety for groundwater.

In the case of small drainage areas and rainwater, also characterized by a low degree of pollution, infiltration wells may be used (Fig. 6.11). The advantage of infiltration wells is the small area of the terrain, which is required for their location.

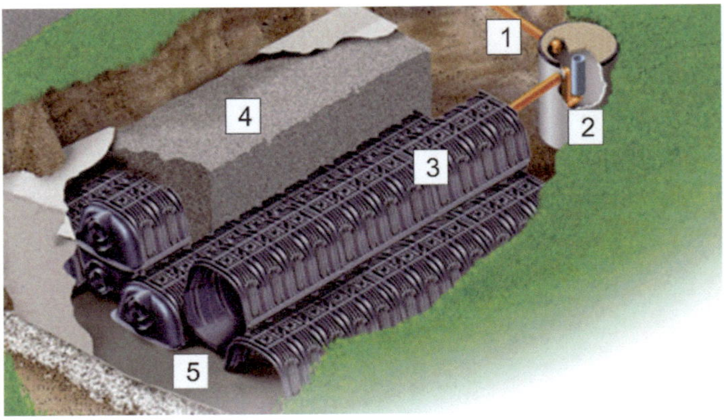

Fig. 6.10 Example of the implementation of the Hauraton infiltration chambers 1—rainwater inflow, 2—treatment device, 3—infiltration chambers, 4—filtration material, 5—nonwoven fabric

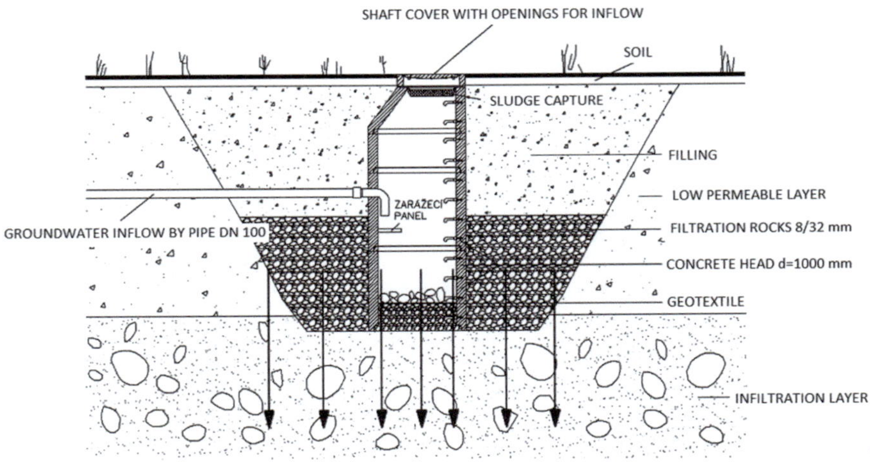

Fig. 6.11 Example of infiltration well

Devices of this type are suitable for the management of rainwater from building plots and roofs of single-family buildings. Infiltration wells are usually made of prefabricated or plastic components. They have an open bottom, and sometimes also perforations in the side walls. Well diameters most often range from 1.0 to 2.0 m, and in the case of a significant amount of rainwater, they are combined into well systems. Due to the low efficiency of water purification in the infiltration well and the need to protect groundwater, the bottom of the well should be located at least 1.5 m above the groundwater level so that rainwater can be treated in the ground.

In the case of areas on which surface rainwater infiltration devices cannot be located, rigoles can be used. They are open or closed underground ditches filled with ground filter material. The material is usually gravel, sand or grit. These materials

Fig. 6.12 Examples of the implementation of rigoles used to drain rainwater from a single-family building roof (photo: Daniel Słyś)

play a structural role as well as create a space for retention of rainwater. This method of rainwater management is characterized by limited capacity to purify it. Therefore, in the case of heavily contaminated rainwater, it is necessary to apply a pretreatment step before introducing rainwater into the rigola or limit the use of such equipment only to water discharged from "clean" surfaces. Properly shaped rigoles can be an interesting landscape element. An example of such a solution is shown in Fig. 6.12.

6.3 Rainwater Retention

The process of outflow of rainwater from urbanized areas is often violent. In such a situation, the amount of rainwater can significantly exceed the hydraulic capacity of sewage systems, devices and objects cooperating with them, and rivers that are their receivers. In order to reduce these problems, rainwater retention facilities are used. Their task is to transform the high instantaneous flow rate into a more favorable flattened hydrogram of their flow with less intensity and longer duration. The method of operation of retention facilities is shown in Fig. 6.13.

It is possible to distinguish the types of facilities and equipment used for retaining rainwater in urban catchments and on sewer networks which were shown on Fig. 6.14.

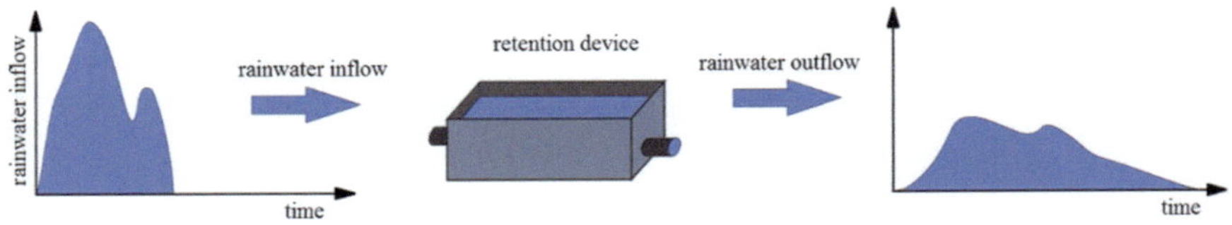

Fig. 6.13 Diagram of the operation of retention facilities

Fig. 6.14 Types of retention devices

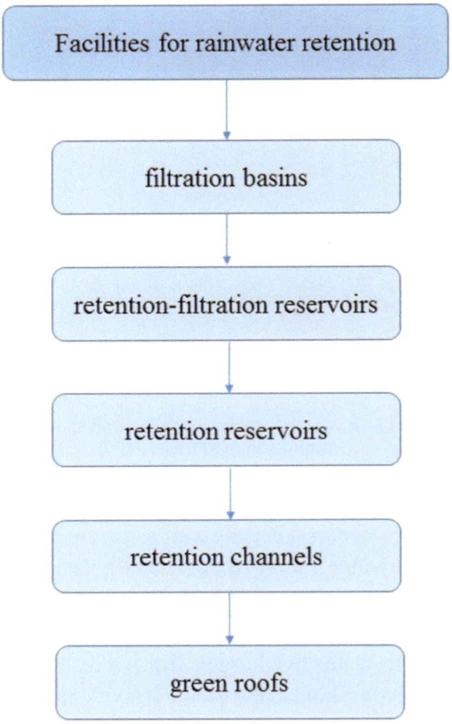

These solutions may have various constructions, applications, rainwater treatment abilities and may perform various functions in urbanized catchments and sewage systems.

In areas where soils with low water permeability exist, the use of infiltration devices is inefficient and financially unprofitable. In this case, a better solution is to use a different structure of facilities and devices to delay rainwater runoff from the catchment, reduce their outflow and additionally allow them to be pretreated.

The retention device with the simplest structure is the filtration basin, which is a similar solution to the infiltration basin. The top layer of the filtration basin is made of fertile soil covered with vegetation. The task of this layer is to purify rainwater during its storage and soaking. Below this layer, there is soil that is a ground filter, in which rainwater is purified in physical, chemical and biological processes. The soil filter should have a large proportion of free space between the grains in which rainwater can be stored. In the bottom part of the soil filter, drainage is performed, whose task is to collect and discharge the filtration water to the collecting well, and then to the receiver. The filtration basin is used when there is a layer of impermeable material below the soil filter and the discharge of rainwater into the ground is not possible and in the case of a high level of ground water. Filtration basin can also be made if rainwater is characterized by a high content of contaminants.

Another solution used for retention and pretreatment of rainwater is a retention–filtration reservoir (Fig. 6.15). It is a similar device for the filtration basin but with larger depths and larger dimensions. Due to the greater filling of the tank, its fence should be used in residential areas. An important design parameter is the time of water retention in the tank, which should be adapted to the type of vegetation growing on its bottom and slopes. The rainwater supplied to the tank slowly seeps through the soil filtration layers. In these layers, physical, chemical and biological purification processes take place. Due to the fact that below the bottom of the tank there is an impermeable ground or artificial seal, the purified water is collected and discharged into sewers or infiltration devices.

Structural solutions of retention reservoirs result from the method of sewage transport and the type of sewage system. The location of the retention reservoir in the catchment should be justified by the assessment of the sewer network capacity, the analysis of the impact on the water quality in the receiver, local physiographic and

Fig. 6.15 Example of a retention–filtration tank (photo: Daniel Słyś)

hydrogeological conditions and land development. Taking into account the location of the reservoir, two main types can be distinguished: surface tanks and underground reservoirs. In turn, due to the hydraulic system used in the tank, retention tanks can be divided into single-chamber (classic) and multichamber, and gravitational, pump-gravitational and vacuum-gravitational. Retention reservoirs can also be divided due to their location in the sewerage system as follows:

- network tanks, hydraulically unloading sewage network and objects cooperating with it;
- averaging tanks, whose task is to stabilize the inflow of sewage to the treatment plant;
- storm tanks, located in front of storm water overflows or on storm sewers, whose task is to reduce the volume of sewage discharged to receivers.

Surface tanks are used as rainwater storage facilities that have not been mixed with domestic and industrial sewage. Therefore, this type of object is intended for waters contamination mainly with mineral and dissolved pollutants, which do not constitute a sanitary hazard. Surface tanks are usually made as single-chamber objects. An example of a solution for such a tank is shown in Fig. 6.16.

Traditional solutions are large constructions, whose location in urban areas may encounter difficulties caused by the lack of sufficient space for the construction of the reservoir. In such cases, it may be a beneficial solution to store rainwater in retention channels made of pipe sections with significant diameters. An example of such a solution may be patent No. 217405 (Słyś and Dziopak 2009). It is a construction

Fig. 6.16 Surface retention reservoir used to drain the road (photo: Daniel Słyś)

Fig. 6.17 Example of retention canal (Adopted from Uponor Infra)

characterized by the fact that the internal space of the canal is divided into chambers by partitions, which are located in canal perpendicularly to wastewater flow (Fig. 6.17). These partitions have flow hole in the lower zone, enabling the transport of sewage to subsequent chambers of the canal.

Green roofs have been used for many years as one of the methods of rainwater management in urbanized areas, since the vegetation and roof layers have high water retention capacity (Chen et al. 2015; Stovin et al. 2012) (Fig. 6.18).

Some water also evaporates, which contributes to improving the local climate. As a result of these processes, rainwater runoff is reduced and delayed to sewage systems, thanks to which they do not undergo hydraulic overloading. In addition, the infiltration of rainwater on green roofs is an important element of flood protection, especially in the aspect of the so-called urban floods. The reduction of water runoff depends on many factors, which include: vegetation type, substrate thickness, type of drainage material, rainfall intensity and slope of the green roof (Shafique et al. 2018). Depending on the design of the green roof, it is possible to reduce the outflow of rainwater even up to 90% (Carter and Jacksson 2007).

Fig. 6.18 Retention green roof (Optigrün)

References

Carter TL, Jackson CR (2007) Vegetated roofs for storm water management at multiple spatial scales. Landscape Urban Plan 80:84–94

Chen X-P, Huang P, Zhou Z-X, Gao C (2015) A review of green roof performance towards management of roof runoff. Chin J Appl Ecol 26:2581–2590

Shafique M, Kim R, Rafiq M (2018) Green roof benefits, opportunities and challenges—a review. Renew Sustain Energy Rev 90:757–773

Słyś D, Dziopak J (2009) Retention sewage canal. Patent no. 217405

Stovin V, Vesuviano G, Kasmin H (2012) The hydrological performance of a green roof test bed under UK climatic conditions. J Hydrol 414:148–161

Chapter 7
Conclusion

Abstract This chapter presents conclusions and benefits from the research of rainwater infiltration for the practical as well as scientifical purposes. The further research in the field of rainwater management in the urban areas is also outlined.

In recent years, progress has been made in the science as well as the practice of urban hydrology in terms of basic understanding and approaches to management. New technologies for recording, analyzing and predicting precipitation in urban areas have been developed to highlight the problems of small spatial and time rates of precipitation and runoff in cities as well as their absorption from surface runoff, with emphasis on the retention capacity of the selected area and the intensity of precipitation.

Rainwater management should be considered as a sustainable strategy for rural and urban regeneration in its environmental, economic and social aspects (Vranyová et al. 2011). In the case of precipitation water infiltration, the interpretation of the law is unambiguous: It is about groundwater treatment, i.e., artificial enrichment of groundwater resources with surface water, for which permission of the relevant authority is necessary. With the development of construction in and around the big cities, there is a growing problem of overloading the sewer network (Zeleňáková et al. 2017, 2019). At present, great emphasis is placed on removing rainwater from the urban and rural landscapes through tapping facilities, which is a sustainable way of dealing with precipitation waters from the surface runoff. For that reason, this dissertation focuses on the experimental and numerical analysis of the solution to precipitation water absorption from the surface runoff. The research focused on determining the filtration coefficient with which the existing tapping equipment was used, the absorption shafts in model location I and the newly implemented acquisition gallery in model location II, based on the correlation of theoretical and practical analysis and synthesis of acquired knowledge in the conditions of eastern Slovakia. The work points out the benefits of using pumping devices, which represent an effective way of draining rainwater from the surface runoff for a particular project proposal.

Conclusions and benefits for the further development of science and the direction of further research are as follows.

M. Zeleňáková et al., *Rainwater Infiltration in Urban Areas*,
Water Science and Technology Library 89,
https://doi.org/10.1007/978-3-030-34698-0_7

The research outputs include verification of the accumulation capacity of precipitation water in the infiltration facilities in the model areas, determination of the coefficients of filtration based on local conditions and their confrontation with the standard requirements, and verification of the table and measured precipitation sums and rainfall intensity.

The benefits in this area can be summarized in the following points:

- theoretical analysis of solved problems and subsequently the use of numerical calculations under experimental conditions;
- field measurements of precipitation sums and monitoring the relative intensity of rainfall for the relevant periodicities, thereby confirming the current meteorological tables;
- experimental rainwater absorption analysis aimed at processing the necessary values for determining the hydrodynamic test filtration coefficient, as well as laboratory tests and calculations from empirical formulas;
- numerical analysis of measured values related to the accumulation of precipitation water, with emphasis on retention capacity of model areas and filtration of surface water outflow;
- creating a mathematical–physical model for predicting the emptying time of water in an infiltration device;
- evaluation and selection of the optimal solution for a rainwater infiltration facility.

Further research in the field of water intake from the surface runoff, with emphasis on the retention capacity of the selected area and the intensity of precipitation, should be oriented toward:

- legislative support in the field of project design of equipment for water absorption from surface runoff;
- preparation of meteorological tables for the occurrence of extreme precipitation;
- events;
- research to monitor correlation of rainfall, precipitation and theoretical imputation results with practical results.

Conclusions and benefits for professional practice

The conclusions and benefits for professional practice include

- assessing the current state of knowledge of the area concerned and the possibility of applying sustainable methods for the management of rainwater;
- own monitoring of precipitation sums and rainfall, confirmed by the current meteorological tables from SHMI Košice;
- comparison of filtration coefficient determination procedures for project intentions;
- evaluating and verifying alternative solutions of the infiltration equipment, finding inadequate input parameters when designing the infiltration device in an inconsistent alternative;
- emphasizing the importance of engineering geological research in designing not only intake facilities but also the construction of buildings;

- comparison of current legislation in the area concerned, pointing out differences and the need to prepare a Slovak technical standard for the design of tapping equipment;
- recommendations for the design of tapping devices based on theoretical analysis as well as experimental research and numerical analysis.

The variability and complexity of rainfall response in cities remain an area of active research. Especially in suburban areas with a complex mix of urban and rural landscapes, there are multiple collision events which complicate urban drainage. While drainage from impermeable areas is relatively predictable, changes in permeable surfaces result in changes in surface and subsurface runoff processes. For these reasons, further research and solutions for the infiltration of surface water are still required.

References

Vranayová Z, Karelová Z, Očipová D (2011) Precipitation monitoring methodology at the Campus of Technical University of Košice. Sel Sci Pap J Civ Eng 6:2

Zeleňáková M, Hudáková G, Tometz L, Hlavatá H (2017) Investigation of rainwater infiltration with emphasis on hydrogeological as well as hydrological conditions In: ICEE 2017. Vilnius Gediminas Technical University: VGTU Press, pp 1–5

Zeleňáková M, Hluštík P, Markovič G, Hudáková G, Tometz L (2019) Percolation of Water from Surface Runoff—case studies. New Trends in urban drainage modelling: UDM 2018. Cham (Švajčiarsko), Springer Nature Switzerland, pp 809–813

MIX
Papier aus verantwortungsvollen Quellen
Paper from responsible sources
FSC® C105338

If you have any concerns about our products,
you can contact us on
ProductSafety@springernature.com

In case Publisher is established outside the EU,
the EU authorized representative is:
Springer Nature Customer Service Center GmbH
Europaplatz 3, 69115 Heidelberg, Germany

Printed by Libri Plureos GmbH
in Hamburg, Germany